Lecture Notes in Electrical Engineering

Volume 94

Vladimir Popovskij, Alexander Barkalov,
Larysa Titarenko

Control and Adaptation in Telecommunication Systems

Mathematical Foundations

Springer

Professor Vladimir Popovskij
Kharkov National University of Radioelectronics
Lenin Avenue, 14
Kharkov 61166
Ukraine
E-mail: tkc@kture.kharkov.ua
http://www.kture.kharkov.ua/

Professor Alexander Barkalov
University of Zielona Gora
Institute of Informatics and Electronics
Podgorna Street 50
65-246 Zielona Gora
Poland
E-mail:
A.Barkalov@iie.uz.zgora.pl
http://www.iie.uz.zgora.pl

Dr. Larysa Titarenko
University of Zielona Gora
Institute of Informatics and Electronics
Podgorna Street 50
65-246 Zielona Gora
Poland
E-mail:
L.Titarenko@iie.uz.zgora.pl
http://www.iie.uz.zgora.pl

ISBN 978-3-642-20613-9 e-ISBN 978-3-642-20614-6

DOI 10.1007/978-3-642-20614-6

Lecture Notes in Electrical Engineering ISSN 1876-1100

Library of Congress Control Number: 2011927169

Typeset & Coverdesign: Scientific Publishing Services Pvt. Ltd., Chennai, India.

Printed on acid-free paper

9 8 7 6 5 4 3 2 1

springer.com

Contents

Abbreviations

AAA	adaptive antenna array
ACS	access control system
AIC	adaptive interference compensator
ALN	adaptive linear neuron
A-RACF	Access Resource And Admission Control Function
BI	business intelligence
BM	Boltzmann machine
BML	Business Management Layer
BPM	business process management
CAC	Call Admission Control
CC	cloud computing
CBQ	class-based queuing
CDMA	code division multiple access
CMIP	Common Management Information Protocol
CPN	coloured Petri net
CWTA	conscience winner takes all
DBMS	database management system
DCN	data-communication network
DM	data manager
DPC	data processing center
DSS	decision support system
EDA	event-driven architecture
EMC	electromagnetic compatibility
EML	Element Management Layer
FAB	fulfilment, assurance and billing
ETSI	European Telecommunications Standard Institute
FSM	finite state machine
GII	global informational infrastructure
IaaS	Information - as - a - Service
IP	Internet protocol
ISO	International Organization for Standardization

IT	information technology
ITSM	IT service management
KBF	Kalman-Busy filter
ITU	International Telecommunications Union
LAN	local area network
LSA	link state advertisement
LTR	logical-transformational rules
MAN	metropolitan area network
MIB	Management Information Base
MIMO	multi-input multi-output
MLC	maximum likelihood criterion
MPC	maximum probability criterion
MOF	Microsoft Operations Framework
MS	management system
MMSD	minimum mean-square deviation
NACF	network addition control function
NGN	Next Generation Networks
NEL	Network Element Layer
NML	Network Management Layer
NMS	network management system
NN	neural networks
NOS	network operating system
OSPF	open shortest path first protocol
PBNM	policy-based network management
PMD	people making decisions
PN	Petri net
PSR	policy system rules
QoS	quality of service
RACF	Resource and Admission Control Function
RACS	Resource and Admission Control Sub-System
RED	Random Early Detection
RM	Robbins-Monro procedure
RMLP	recurrent multilayer perceptron
RMON	Remote Network Management Protocol Information Base
RSVP	resource reservation protocol
SCF	service control function
SLA	Service Level Agreement
SML	Service Management Layer
SNMP	Simple Network Management Protocol
SOA	service-oriented architecture
SOFM	self-organizing feature map
SPDF	Service-Based Policy Decision Function
TCP	transmission control protocol
TCS	telecommunication systems
TM	Turing machine

TMN	telecommunication management network
TSP	travelling salesman problem
VQ	vector quantization
WFQ	weighted fair queuing
WAN	wide area network
WTA	winner takes all
WV	weight vector

Chapter 1
Introduction

Intensive development of digital technologies coincided in time with the beginning of the new era in telecommunications. It made possible to formalize many procedures of data exchange and to atomize some operations which made providing of service and make work of many telecommunication workers much easier. Some new telecommunication technologies were born out of the necessity for use of specific configurations of network elements and networks, as well as for a possibility of providing maximum characteristics of efficiency combined with high requirements to the stability of operation, the overcoming of different catastrophic situations and deadlock conditions, such as failures and "pending" of the network and the like. The threshold between information systems and telecommunication systems has become practically invisible. It resulted in such a new term as "infocommunication".

Efficient functioning and further development of infocommunication continue right along. A modern stage of development and perfection for both information and telecommunication systems is carried out due to technological innovations and customer demand for services. Unfortunately, there is no exact scientific approach used for development of infocommunication. It is possible to use the system theory for designing infocommunication systems, but such a possibility is only declared.

There are few reasons of such a weak usage of the results of system theory for development of infocommunication. Firstly, the evolution of the system theory has been slowing down from the middle of the XX century because of the lack for evident need from the side of applied problems. Secondly, new and new technologies used in infocommunication allow rapid evolution and introduction of infocommunication without obvious need in use of a theory. And finally, numerous developers and manufactures of the new networking equipment and, therefore, technologies take interest rather in finding a decent niche of the telecommunications market, rather than in systematizing and unification of their products.

In spite of these objective and subjective reasons, there arises necessity to use system theories, especially cybernetics, which contributed greatly to solving problems related to infocommunication. Now, different cybernetic approaches are used in this area, where a telecommunication system (or its part) is viewed as a dynamically

V. Popovskij et al.: Control and Adaptation in Telecommunication Systems, LNEE 94, pp. 1–4, 2011.
springerlink.com

controlled system. It allows application of formalized differential models operating in accordance with some criteria, which can be stochastic, as well as methods of overcoming of a-priori uncertainty together with methods of decision makings, assessments, extrapolation, interpolation and control.

The powerful apparatus of cybernetic solutions is successfully used in the one-dimensional variant. A lot of one-dimensional methods and procedures have been thoroughly investigated. At the same time, there are a lot of problems in the cases of multi-dimensional models and representations avoided by many authors. These problems are connected not only with the necessity of justification of Markov's properties of a multi-dimensional model, but also with algorithmic solutions and interpretations of gains and losses considering the existing interrelations among the components of a multi-dimensional system. As it is obvious, namely these interrelations make the system. It disintegrates on separate elements without interrelations among the components. Due to these interrelations, a system gains new properties, which are not the sum of the properties for its components. Such an effect is known as emergence.

Therefore, on the one hand, there is a necessity for use of the results of the systems theory and cybernetics in existed infocommunication technologies. It allows usage of leading technologies on the base of strict mathematical criteria and background. On the other hand, the apparatus of the systems theory should be adapted to the reality of infocommunication systems. Therefore, the existed theory should be concretized to the specific of infocommunication.

A vast majority of existed algorithms require some theoretical foundation. These are the algorithms of signal processing, control and adaptation either used or waiting for use in existed technologies. In addition, the infocommunication systems became more and more similar to the systems of controlled automata. It requires some constructive theory which can explain the operation of control subsystems of the infocommunication systems.

Our book is written on the base of statistical probabilistic approach. From our point of view, it is very important because it permits to describe a wide class of situations. It makes our approach different from the deterministic one, when a task results in a single solution. Our approach leads to more stable solutions under the conditions of random and non-stationary processes, traffics, signals and interferences. Special attention is spared to problems of preparation of statistics, because it influences the final objective conclusions.

Mathematical models of systems, processes and functions are based on the methods of state variables, providing possibility for adequate presentation of these models as some dynamic objects. These models are used to construct estimation algorithms known as filters of Kalman-Busy, Robbins-Monro and so on. Just the procedures of estimation take the central place in the problems of control for stochastic systems in common and telecommunication systems particularly. The main role of estimations follows from the fact that they permit obtaining specific values rather than general presentations such as probability distributions. These specific values can be used for further control, for finding other solutions, for constructing adaptive procedures, or identification of unknown models.

We tried to keep unity of designations and terms. But in the case of some specific topics (such as the automata theory, or the theory of neural networks) we kept designations and terms which are traditionally used in these areas.

The monograph includes ten chapters. The short content of these chapters can be found bellow.

Chapter 2 is devoted to the background of controlled systems. The properties of models are discussed for complex controlled systems, as well as their classification, the peculiarities of situational and automatic control methods, the rules of system policy, the tasks and functions of controlled systems. The network of telecommunication control TMN is used as a basic methodology of control.

Chapter 3 is connected with control technologies used under the operation of telecommunication systems (TCS). The specific control technologies and different models of services are discussed oriented of increase of quality of services provided by TCS. This chapter includes the analysis of the technologies for control of resources, structure and functional states of TCS. Some important properties of network control protocols such as CMIP, SNMP, RMON, Net Flow are discussed.

Chapter 4 is devoted to mathematical models and control principles used in telecommunication systems. Particularly, there are discussed static and dynamic, deterministic and stochastic models of controlled systems and decision-making techniques of control modes. The methods of state variables are used to construct these models. Some criteria of control optimality are discussed, such as the compatibility criterion, leading to the principles of guaranteed quality, different criteria of optimality and preference criterion. The criterion of minimum of square deviation is discussed thoroughly. Both the algorithm of optimal control and main principles of control systems' construction are considered. These principles are based on methods of Ponselle and Watt. The decomposition theorem is formulated.

Chapter 5 is devoted to the methods of providing controllability. Some conditions are discussed, such as observability, identifiability, stability, and invariance. Attention is paid to the methods of sample statistics under the assumption of observation of random variables, random processes and random fields, as well as sample parameters for different values of correlation windows. The recommendations are given for construction of sample and recursive estimates, used for different random objects, as well as for different criteria of their optimality.

Chapter 6 deals with methods used for synthesis of algorithms of recursive estimations, such as the procedure of Kalman-Busy, Robbins-Monro and so on. These algorithms are implemented in both analogue and digital variants. The peculiarities of recursive calculations of estimations are analyzed. The recommendations are given for providing stable operation modes of functioning for estimation procedures. There are given results of investigations of sensibility of Kalman-Busy filter to deviations of chosen model from the real situation.

Chapter 7 is devoted to synthesis of control algorithms. Two main approaches are discussed, namely, the control of the state and the control of the observation. Both approaches are considered from the point of view of the method of state variables. The examples are given for solution of the tasks of state control under fulfilment of conditions for the decomposition theorem. The problem of observation control is formulated as a task of control of the observation basis. This control should provide some necessary properties of the observed useful signal, such as its interference protection. The last class of problems is reduced to construction of adaptive compensator of interferences and adaptive antenna arrays, as well as to the tasks of spatial-temporal encoding and spatial-temporal access to the base station in the system of mobile communications.

Chapter 8 discusses the topic of taught controlled systems. Two kinds of taught systems are considered. First of them is a learning by instruction, whereas the second has no teacher (instruction). There are some examples of practical implementation of these approaches in the existed telecommunication systems. The following classes of taught systems are analysed: the systems with identification of a model, the systems of the search type, and the self-organized and self-repairing systems with re-engineering. The entropic, homeostatic and morphogenetic solutions are discussed.

Chapter 9 is devoted to neural networks and their application into the control tasks. The connection of perceptron with the tasks of observation control is discussed. The peculiarities of functioning are considered for neural networks with different organization, as well as self-organized and self-taught networks. Few examples are given for practical solutions, such as the travelling salesman problem and Boltzmann machine.

Chapter 10 is devoted to discussion of the theory and methods of multifunctional control automata. The methods of Petri nets and E-nets are also discussed. These methods are used for simulation, analysis and development of telecommunication networks.

Chapter 11 is devoted to management of business processes (BPM). Its role for telecommunications is analysed. In the same time, the role of infocommunication for organization and management of business processes of the general nature is discussed. Some other problems are discussed here, such as the processes of management of activity, peculiarities and main directions of development of infrastructure of information systems, the role of new system technologies of virtualization, service-oriented architecture (SOA), grid-mechanisms, and so on. The restrictions are analysed springing up under the management of business-processes.

Our book targets on students, PhD students and professionals in the area of telecommunications. We hope it will be useful for everybody connected with the new information technologies.

Chapter 2
General Information about Controlled Systems

Abstract. This chapter is devoted to the background of controlled systems. The properties of models are discussed for complex controlled systems, as well as their classification, the peculiarities of situational and automatic control methods, the rules of system policy, the tasks and functions of controlled systems. The network of telecommunication control TMN is used as a basic methodology of control.

2.1 Introduction to Telecommunications, Their Models and Situational Control

Now, we witness the wide usage of information practically in all areas of human activity. The global informational infrastructure (GII) is creating, allowing any consumer fast information exchange in any time and any place of the world. The telecommunication systems (TCS) are assigned for providing data exchange with required quality. In this sense, telecommunication system are infrastructures of information systems. In the same time, TCS solves some very important self-reliant functions connected with providing customers by data services. Moreover, it is very difficult to distinguish telecommunications and information systems. Such a situation leads to coin of the new term "infocommunications". These systems are advanced in unbelievable fast pace. Their evolution is provided by introduction of new and new technological and technical achievements. In the same time the following paradoxical fact takes place: till now, there is no the theory of telecommunication systems.

Obviously, such a theory should be based on the general system theory. From the system approach's point of view, the most adequate mathematical model of a telecommunication system is represented as a complex distributed controlled logistical structure, processing some random traffic. With help of the control under the system and modes of operation for its individual elements, it is possible to use the network resources in the best possible way. It allows a real-time reaction on any changes of system traffic and user requests, as well as providing stable operation of

V. Popovskij et al.: Control and Adaptation in Telecommunication Systems, LNEE 94, pp. 5–17, 2011.
springerlink.com

the system in both real and perspective time together with maximum profitability of the system.

The uncontrolled communications system used in the past can provide connections only between objects 1 and 1′, 2 and 2′ and so on (Fig. 2.1). Here symbol M/D stands for multiplexer/demulteplexer, whereas symbols 1,2,. . . ,n determine some terminals of the system. In such a system, its resources are used in very inefficient way. Adding of a commutator in this system makes it more flexible, because the commutator (switch) gives to a user the opportunity of connection with any other user (terminal) of this system. It expanded significantly the facilities of the system.

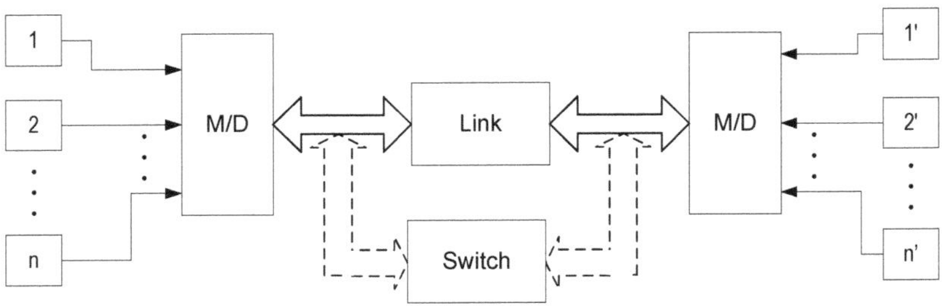

Fig. 2.1 Structure of uncontrolled communications

The telecommunication systems with batched (packet) transmission possess even more facilities for use of network resources. These networks operate, for example, using the Internet protocol (IP). A modern telecommunication system can be represented as a very complex, distributed through large territory, automaton, providing moving and delivering information from its source to a receiver. This automaton operates under the influence of preliminary loaded programs and algorithms. Its operation depends strongly on some devices using to correct its modes of operation and structure of interconnections, as well as on the will of human beings making solutions, which influence the mode of network's operation. All these components can be viewed as a control complex of communication network. Efficient use of network resources is achieved due to automatic delivery of the packages (packets) to an addressee, choosing optimal routes of packets' transferring, and control of the quality of services offered by a network.

A conception of control is very wide. There are two terms used for it, namely, control and management. Orders of a chief of an organization are treated as a control, as well as some orders of people making decisions (PMD). Such kind of control is named declarative. The system of connections is shown in Fig. 2.2 providing different control methods inside a telecommunication network. It is taken into account that a final decision belongs to a human being (PMD), despite any level of automation.

Therefore, there are many possible control modes which can be used in a telecommunication system. A possibility for implementing any type of control is based on representation of control modes as some models. A distinguished feature of complex systems, including telecommunication networks, is absence of a single mathematical

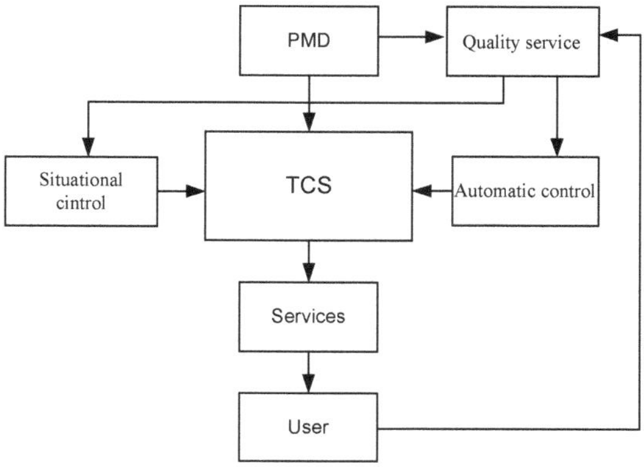

Fig. 2.2 Connections of different control methods in telecommunication systems

model for their representation. These systems are represented by a wide variety of different models. Each of models, in correspondence with the system theory, reflects some property of a system, such as: general system properties (integrity, stability, observability, controllability, openness, dynamics, reliability, and so on); structural properties (composition, connectivity, complexness, hierarchy, scalability, and so on); functional properties (persistence, performance, efficiency, accuracy, economy, and so on).

The structure of classifier of the properties of TCS is shown in Fig.2.3. Such big amount of models in use is a result of the complexness and bulkiness of general representation, as well as with the fact of lack of possibility to formalize many processes running inside complex systems. It makes very difficult both analysis and synthesis of mathematical models, as well as the reasonable choice of adequate control criteria. As a rule, in such cases they use the methods of situational control for a given complex system. Here, the final decision belongs to PMD in charge. Therefore, a human being makes a decision about the general control of a system (or its part). To do it, PMD uses his experience, intuition, possible automation of some procedures, operations and processes, and some system for support of decision-making (DSS, decision support system). Such kind of control is named the situational control in contrast to the formalized control based on the rigorous automatic control theory.

The core of situational control methods is some semiotic model. A model is named semiotic if it is represented using elements of a language used by PMD. As a rule, it is a collection of audio-visual signs, signals and pictures, showing the current and, maybe, next situation of a system. The structure of semiotic control model is shown in Fig. 2.4.

There are some peculiarities of the situational control, which make it different from other methods of system control. They are analysed bellow.

1. The situational control requires many efforts for creating the preliminary data base with information about an object of control, the rules of its operation and approaches for their controlling. These efforts are justified only in the case when it is

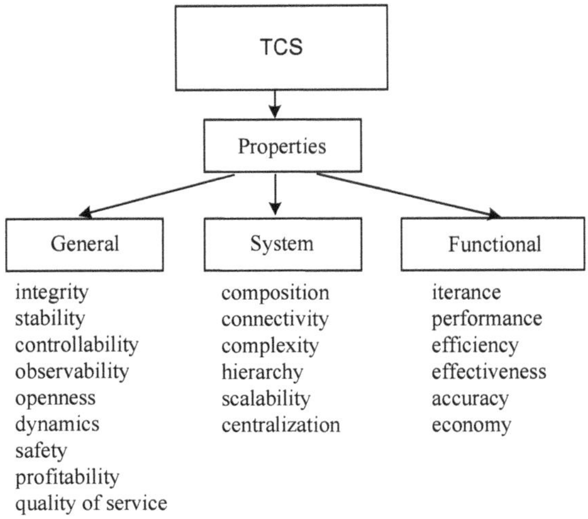

Fig. 2.3 Classification of properties of TCS

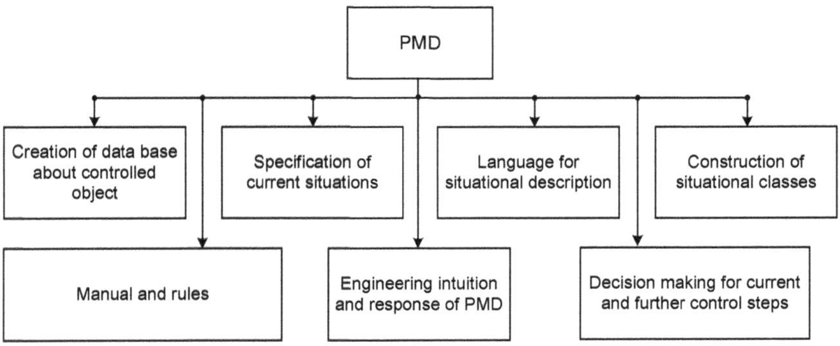

Fig. 2.4 Semiotic model of control

impossible to use other known methods for specification of the object, as well as procedures of its control. For example, there is no need in the situational control if an object can be represented by a system of differential equations. But the situational control can be implied in the case when there are thousands of equations in such a system.

2. It is very important to make the correct choice of a language used for specification of current situations in the object of control. Such a language should reflect all main parameters and connections, needed for classification of the specifications and making a single-step control decision. In this case, it is very important to make the proper choice of the level of specification. If it is too detailed, it leads to the "noise effect", when some minor particulars make it difficult to understand the real situation. If it is not thorough enough, it complicates the process of decision making, too.

3. A situation definition language should reflect not only all numerical facts and relations used to characterize the controlled object, but also some qualitative knowledge which cannot be formalized from the mathematical point of view. In the overwhelming majority of situations, the PMD can receive only inexact information from some process engineer. For example, "If the condition A takes place, then I think it is better to start the process B" or "It seems that if A increases, then B will be decreased". Sometimes, it is very difficult to formalize similar statements. But to make decisions, these statements should be expressed using some language of situations' definitions. It should be taken into account that statements of a human being about a controlled object are very often far from completeness. It means the special approaches should be used for extracting all necessary information from a technologist, for example.

4. The classification of situations and their combining into some classes under the use of single-step control decisions is executed on the subjective base. It is connected with the fact that any primary information about any current situation and corresponding control decision is taken from experts. A system (either automated system or PMD) should summarize knowledge received from different experts. It makes the system a carrier of collective (joint) experience of many different people. But the classification procedures should be built in such a manner that the classification can be used for those current situations where there are no recommendations from experts. It results in reducing the classification problem to a problem of conception construction on the base of learning sequences. When a system constructs some conception, it possesses a bigger amount of knowledge in comparison with the conceptions obtained from experts in the beginning. Of course, this additional knowledge can turn out to be incorrect. This fact can be revealed in the process of system's operation. Therefore, some undesirable things can appeared in a system, such as "strange situations", incorrect conceptions, and wrong generalizations.

5. Instructions and rules used for situational control play very important parts for a criterion of decision making. These rules are named logical-transformational rules (LTR). In the beginning, they are formed on the base of information obtained from experts. These rules are refined in the process of the system's operation. It can be done by elimination of contradictions, which leads to construction of new rules. It is correct for extrapolation rules, as well as for estimation of this or that current situation.

6. It follows from analysis of the last two points, that systems of situational control cannot be targeted on the process of optimal control. They are oriented only on such a control, when results are not worse than in the case of human control. But the practice of these systems' application shows that they produce results which are better than the ones obtained under the control by an operator. There are some reasons, which can explain this phenomenon. First of all, these systems are free from influence of human emotions and, therefore, they make correct decisions in both normal and critical situations. Next, the system never forgets anything and takes everything into account to make the best decision. Thus, the method of situational control can be viewed as a heuristic one.

7. Indeed, one-step decisions do not determine the control strategy for majority of real controlled objects. It these objects, it is necessary to form a chain of one-step

recursive decisions to get a final control decision. To do it, the extrapolation system should include some procedures for "concatenation" of one-step decisions. It allows construction more and more complex control decisions.

Taking all above mentioned into account, we can state that some situational heuristic criterion of suitability of obtained results should be used in the systems of situational control. For telecommunication systems, such a criterion provides both quantitative and qualitative estimate for used control approach.

There are different classes of the modern telecommunication systems such as local area network (LAN), metropolitan area network (MAN), wide area network (WAN). Their analysis shows that they belong to the class of systems with situational control, executed on the uppermost layer of their hierarchical structure. In these systems, some control problems are executed as formalized (for example, routing, overload prevention and so on). But these problems are situational too, because they are not optimized and they are solved using some programs, which can be viewed as a sequence of actions caused by some situation. But in modern telecommunication systems, there is a clear tendency for execution more and more procedures in the automatic way. Obviously, the tendency will be kept for increasing specific weight of automatic control methods in TCS. The peculiarity of optimal control is not only obtaining the best solutions for the shortest possible time, but, that is more important, the ability for reaction on the wide class of specific situations (not only reaction on a single specific situation).

Therefore, modern telecommunication systems need modernization of control methods in use. The most important feature of these methods is automation of control procedures. It permits to transfer from simple one-step methods to really complex multi-step control. Lately, a lot of modernizations have appeared in telecommunications technologies. They target providing more general and qualitative observability, monitoring and control and can be viewed as a network management system (NMS). These innovations are oriented also on optimization of control procedures for routing, for example, as well as for some specific functions for specific network elements. As an example, two modern monitoring systems are introduced into telecommunication systems right now. They are the systems named RMON and Net Flow. Both systems are some extensions of the protocol SNMP. They can be used for constructing more general problems of optimal control.

2.2 Methods of Optimal Automatic Control, Rules of System Politics and Tasks Solved by Control

There are a lot of different theories and methods of optimal control. Very often, such tasks are included in the class of optimal control's problems as usual optimization tasks of linear or nonlinear programming. From our point of view, the problem of optimal control is more general in comparison with strictly optimization tasks. The control problems are more dynamic; as a rule, they include the problems of monitoring, identification and so on (Fig. 2.5). On the other hand, the control problems are very specific and oriented on dynamic applied decisions having some restrictions.

It means, the control problems are more special. Therefore, the control problems include the optimization tasks as their parts.

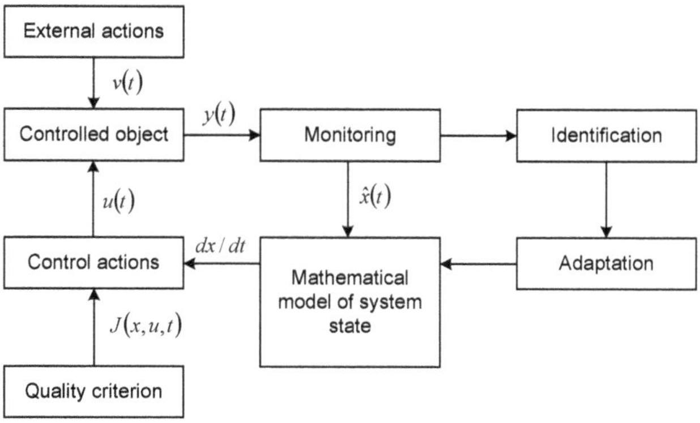

Fig. 2.5 Structure of functional connections in the problem of optimal control

There are three original positions in the standard representation of an optimal control's problem (Fig. 2.6): a mathematical model of a system $S(x,u,t)$to be optimized; an optimality criterion concerned either structure, or operating of a automated system, or both, $J(x,u,t)$; contingencies (restrictions) such as $x \in X$, $u \in U$, $t \in T$ and so on, determining the area of possible solutions.

Our book zeroes in considering namely the automatic control methods which are viewed now as an alternative for the situational control methods. The control methods are oriented on stochastic systems, because, as a rule, a telecommunication system is a stochastic system due to its random traffic. Additionally, there are a lot of random factors operating in the system; they can be both internal and external. The practical interest in stochastic systems is explained by the fact that the stochastic solutions are oriented on the wide class of possible situations (in contrast to deterministic solutions).

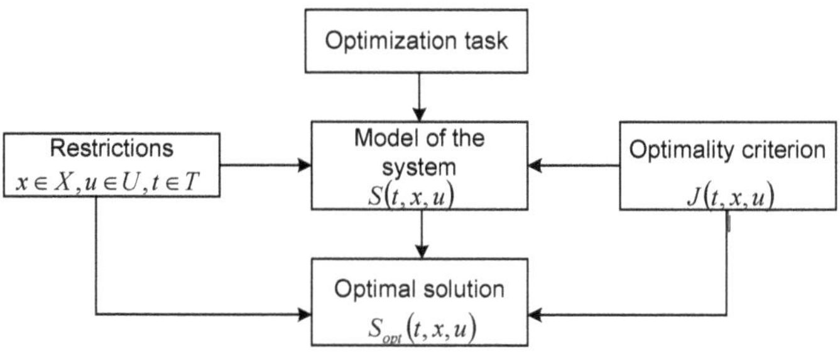

Fig. 2.6 Schematic solution of optimization task

In the beginning of digital TCS and creation of modern telecommunication technologies, some control approaches were either inefficient or contradicted to their functions of the system. For example, the simple control protocol SNMP takes away a significant part of a network resource without giving solution of the final control task. Another example, the standard routing procedure was applied in telecommunication systems and it was based on finding the shortest paths between the objects. But this procedure used only near 30% of possible data throughput due to utilization of these shortest paths (the other, adjacent paths were not used in this case).

These examples illustrate a lack of system approach for solving this or that specific optimization task. The system approach is not a simple rule of action or a specific algorithm. It can be rather viewed as a principle of operation, as a methodology for solving specific problems in the scale of some system. The core of the system approach is the correlation of any specific solution applied to a part of the system with objective (efficiency function) and main interests of the total system.

Because of it, the IETF committee proposed the rules of system policy. These rules are summarized in the special report REC 3198 and are named as policy-based network management (PBNM). The essence of these rules is a necessity in using the system approach for any changes in TCS (or their control). These rules propose three main hierarchical levels (Fig. 2.7): reconfiguration of the network on the base of PBNM (policy-based configuration); configuration of PBNM rules (configuration of policy); configuration of the solution policy on the base of PBNM (policy-based configuration of policy).

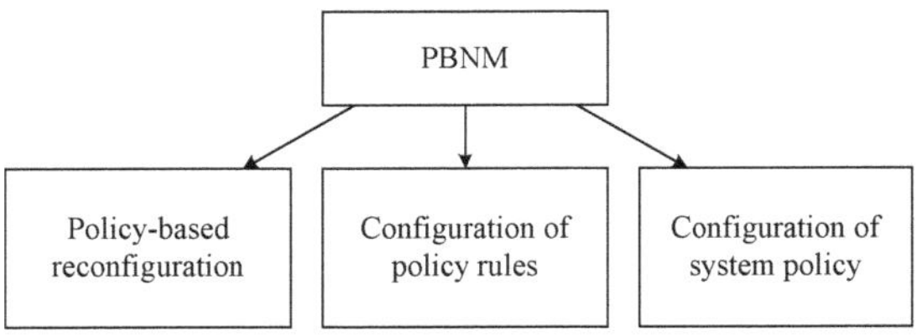

Fig. 2.7 Components of system policy

In this case a property of integrity (emergence) should be taken into account. It means that each component of a system contributes in formation of general properties of the system. It means, too, that changing properties for any component can result in alteration of system properties and, as an extreme point, in its failure.

It is accepted in the system theory to consider so called task-oriented (dedicated) systems. Telecommunication systems belong to this class, because they process traffic and offer services with required quality level. The task-oriented systems can solve the following classes of problems (Fig. 2.8):

1. Targeting (identification). It is identification of a current state or behavior either a system or its specific components.
2. Restructuring. It is any change of the network structure for fulfillment of some required property (for example, restoration after failure of some network components, or redistribution of network resources).

3. Reconfiguration. It is a change of configuration of data flows under influence of random or non-stationary traffic. This task is executed for control of commutator, router, or gateway.

4. Stabilization. It is a keeping system in some required state under influence of some external perturbation actions. Such a property is named invariance.

5. Alteration of system coordinates. It is any change of a system state. Such a change is executed in the phase space as a movement from one point to a final point. This movement can be executed with a given trajectory or in a given (required) time. In this case the given optimization problem either can be solved as a terminal task for the time t_F or it can become a task of stabilization on the infinite interval.

Obviously, the solution of the tasks of restructuring and reconfiguration is based on the morphogenetic approach, whereas the homeostatic approach is used for solving both stabilization and state alteration problems.

Now, there are some known conceptions proposed for control of TCS. The most known are two conceptions, TMN and TINA. In the same time, recommendations given in these technologies are based on general ideas and principles of control. Anything but all their ideas and principles are worked through in details. For ideas which are worked through in details, this detailing is hidden in the system software (but there is restriction on the access to this software). Let us point out that all these details are based on classical algorithms, which in turn are based on classical theories such as the estimation theory, the theory of optimal control, identification methods, methods of decision making (decision-making techniques) and so on. These methods are discussed in our book.

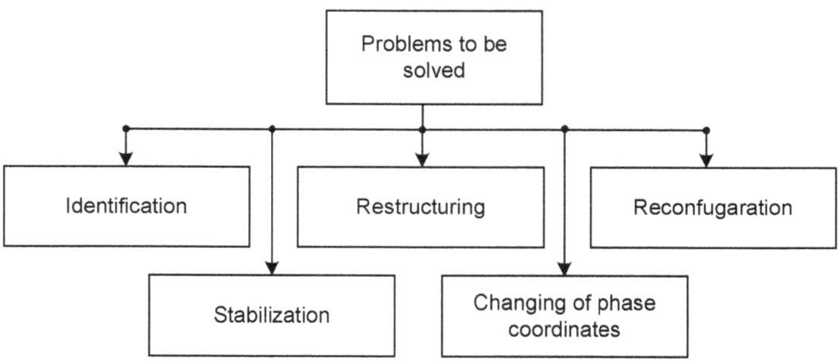

Fig. 2.8 List of problems solved with help of control

2.3 Organization of Control in Telecommunication Systems

Control functions in TCS are various enough and their complete presentation is not obtained yet despite of efforts of designers and producers of these systems. In the case of TCS, there are three main functional groups of problems in the general problem of control: control of infrastructure; control of system quality and provisioning; control of interfaces for communications with clients.

The first problem includes such tasks as planning and development, construction, control of the state of service park, preventive maintenance, and data control. The second problem includes the tasks of service definition and configuration, providing of maintenance, quality control, prices and discounts. The third problem includes sale of services, order processing, reclamation processing, quality assurance, and billings.

We are interested in the control of TCS in tote (the situational control) permitting influence of PMD, as well as the problems of automatic optimal control. The second problems are mostly connected with backbone transport layer.

The methodological base for solving control problems in TCS is the conception of telecommunication management network (TMN). It is oriented on organization of the integrated control of networks with different structures, configurations, levels of traffic, and types of loading. The ideas of TMN are implemented through some separate networks communicated with elements of TCS using unique interfaces and protocols. The control objects for TMN are telecommunication resources representing different network equipment. Exchange of instructions and data among TMN and carriers is executed using some reference points. The network of interconnections of TMN and telecommunication system is shown in Fig. 2.9.

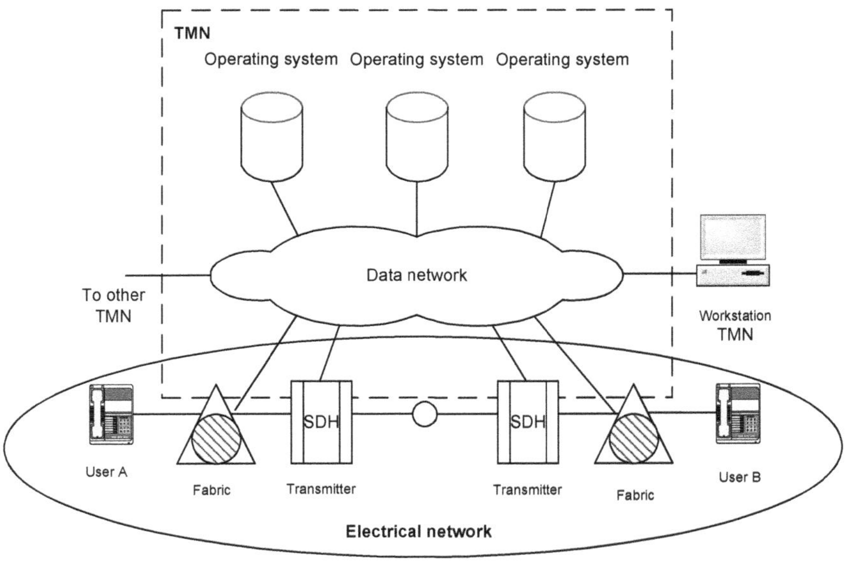

Fig. 2.9 TMN and electrical network

According to the conception of TMN, control functions can be implemented either as some automatic procedures or by an operator (PMD). The network TMN can be provided by one or more operating systems.

Functional abilities of the TMN network can be divided by the following layers: Network Element Layer (NEL); Element Management Layer (EML); Network

Management Layer (NML); Service Management Layer (SML); Business Management Layer (BML). These layers are shown in Fig. 2.10 together with their functional blocks and supporting points.

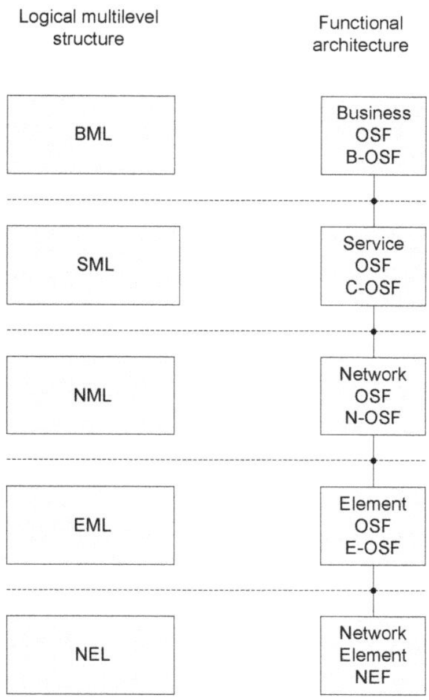

Fig. 2.10 Logical and functional multilayer architecture of TMN

The Network Element Layer is represented by telecommunication equipment, operating under control of special program-agent. It executes information gathering and processing of control signals from the previous layer.

On the Element Management Layer, separate network elements are controlled by the operating system (E-OSF). This layer serves for interconnections with specific functions of the network equipment. The implementation of these functions depends on a producer. As a result, the specific functions are hidden by the EML from other layers of TMN model. The following specific functions can be executed on this layer: error detection (diagnostic) of telecommunication equipment and communication system; measurement of consumed power; measurement of temperature of equipment; measurement of used network resources, such as central processor element's utilization, existing of free place in the input/output buffer, queue length and so on; statistical data logging; modification of software.

It is worth pointing that the operating system for this level and the element itself can be implemented as either separate modules or single hardware-software module.

On the Network Management Layer some different control functions are executed. They concern interrelations among different types of telecommunication equipment. For this layer, the internal structure of network element is "invisible".

It means, for example, that the state of input/output buffer cannot be directly controlled by this layer.

The following functions are executed on this layer: creation of informational model for given network; creation of alternative (by-pass) routes for making connection to support the quality of service (QoS) for end users; modification renovation of routing tables; monitoring of utilization for communication lines and channels; optimization of network facilities for increasing efficiency of utilization for its resources and systems; error detection in the system software.

The operating systems of this layer (OSF) use control data, which do not depend on system manufactures. This information is provided to OSF from the previous layer, NEL. This operating system operates as a manager-program, whereas it is an agent-program for the layer of NEL.

The Service Management Layer deals with control problems concerning directly the users of the network. They are clients of operator, subscribers, and management representatives of operators or providers. The service management is executed on the base of data providing by NML. Let us point out that SMN does not "see" the detailed internal structure of the network. Some subsystems cannot be controlled from this layer, such as routers, automatic switching centres, transmission systems and so on.

Some examples of control functions of this level are the following: quality control of communication services (delays, losses and so on); calculation of volume and utilization of communication services; removal and adding of users; assignment of network addresses and telephone numbers; maintainability of some group of addresses or numbers, for example, for an attached operator.

Formulation and application of the term "service management" is one of the most valuable contributions of TMN conception into the development of control system for services and communication networks. There are a lot of cases where the service management can be used.

The first case: two operators make data exchange of control information to control their interrelated networks (inter-operator control). Obviously, each of them hides the internal structure of his network from the other operator. It is done for the sake of the network safety under conditions of competition on the communication market. The exchange is executed only in the part of control data, which is necessary for providing required services. It could be, for example, some data about preferences of a prescriber or the profile of prescriber's services.

The second case: the first operator provides some communication services using the transport network of the second operator. Such a case is common for providers of services in IP-telephony or other IP-services, which use the network of automatic switching center to connect IP-routers.

The Business Management Layer is responsible for management of the total enterprise. It should be concerned in the very wide context, where the communication control is only one part of this management. The business management should be viewed as some purpose objective rather than the goal achievement. Because of it, the business management is connected rather with the economical aspect in the control strategy for electric communication networks, than with operating control of the network. Some problems of this layer are discussed in Chapter 10.

Using the logical multilayer architecture TMN, it is possible to divide the logical partition of management systems (MS), which can be viewed as a physical implementation of TMN principles. The management systems are either distributed or centralized computer system including servers, workstations and personal computers interconnected through the data-communication network (DCN). The servers and computers are provided by diverse software: network operating systems (NOS), remote access software, data manager (DM), operating systems of workstations, applications for control of electric communication and tools for these applications' control.

Let us discuss the functions of the transport layer of telecommunication network. The main functions of this layer are the following: support of required quality of services; providing independence from both upper and down layer; end-to-end transmission; providing the transparency of transmission; support of addressing for the users of this layer.

The quality of services for this layer is determined by the following parameters: the delay in the path setting; the probability for successful (unsuccessful) ending of the procedure of path setting; throughput; the delay in transit; the error coefficient (probability); the probability of successful (unsuccessful) data transmission; the delay in breaking communication path; the probability of information about the breaking communication path; security; priority in setting agreement about the quality of service among users; the flexibility of transport connection.

The problems of this layer should be solved under: reservation of the frequency band; management of the frequency band; control of routing and planning of routes; application of cache-technologies; queue management.

Being a dynamic system, up-to-day TCS has no opportunity for operating in only single mode or using only single protocol. It is impossible to create such a protocol, because the telecommunication systems are very complex logistical systems and they cannot be represented using only single mathematical model. But a lot of tasks and functions of TCS can be formalized. Some algorithms, rules and protocols should be developed for formalizing these functions. These tools are combined in the mutual system of the network management.

Recommended Literature

1. Ashby, W.: Introduction to Cybernetics. Routledge Kegan & Paul (1964)
2. CISCO. Cisco Networking Essentials, vol. 2. Cisco Systems, Curriculum Development, Team Worldwide Education (1998)
3. Dorf, R., Bishop, R.: Modern control systems, 11th edn. Pearson Prentice Hall, London (2007)
4. Jahne, B.: Digital Image Processing. Springer, Heidelberg (2005)
5. Mesasarovic, M., Takahara, Y.: Theory of Hierarchical Multilevel Systems. Academic Press, London (1970)
6. Mesasarovic, M., Takahara, Y.: General system theory. Academic Press, London (1975)
7. van Gigch, J.: Applied General Systems Theory. In: Harpercollins College Div. (1978)

Chapter 3
Control Technologies in Telecommunication Systems

Abstract. This chapter is connected with control technologies used under the operation of telecommunication systems (TCS). The specific control technologies and different models of services are discussed oriented of increase of quality of services provided by TCS. This chapter includes the analysis of the technologies for control of resources, structure and functional states of TCS. Some important properties of network control protocols such as CMIP, SNMP, RMON, Net Flow are discussed.

3.1 Introduction into Control Technologies for QoS Providing

The implementation of control algorithms for TCS is executed taking into account a lot of specifics. These specific features are the following ones: the state of distribution of the network in both space and time; existence of different delays in the control loop connected with state of distribution, as well as with delays for separate network elements (buffers, operating units, and so on); the principle "Agent/Manager" is used under interaction of separate control objects (applications, network elements); the interaction among network elements is executed using the ordered hierarchic collection of protocols providing sequential processing of control data. One of the most wide used protocols is the Simple Network Management Protocol (SNMP) used for monitoring different network components and carrying monitored data into the control center for visualization of a current situation (state of the system).

It is worth mentioning that there is no some completed control system for telecommunication networks (especially, for such networks as MAN and WAN). Of course, there is the TMN conception, discussed before, but it is very cumbrous to be used in practice. The situation is quite typical when only relatively disconnected independent procedures are used to provide the required quality for services, or to control network resources, structure, functions and so on. Let us add that a majority of existed algorithms are situational, nonoptimal, and rather deterministic. There is the obvious necessity in use of probabilistic approaches because they are

V. Popovskij et al.: Control and Adaptation in Telecommunication Systems, LNEE 94, pp. 19–35, 2011.
springerlink.com

oriented on classes of situations, whereas the deterministic approach targets finding solutions for specific situations.

The communication services given to a user are regulated by some agreements, for example by the standard Rec Y.1540. The following properties should be included in such an agreement: the data throughput (data transfer rate) of a network; the reliability of a network and its components; delays; variations of delays (jitter); loss of packages.

The agreement for service quality is named SLA (Service Level Agreement), sometimes it is called a traffic contract. It is made between a user and a service provider. The agreement contains main features (profiles) of the traffic given by the provider and the parameters of QoS (up to five parameters). The conception of QoS can be defined as an aggregate rate of service characteristics determined a degree of user satisfaction from the given service. As a rule, the practical QoS includes the following characteristics: delay of connection; data throughput; transmission quality.

This agreement between a user and a service provider can be made either before a particular communication session or for some time period. It is possible a situation when the real level of service SLA is lower than required and determined into the agreement. In this case some sanctions are determined in the agreement. For example, an operator should decrease prices or incur a loss. The agreement SLA is used for control of network services of SML.

As follows from analysis of the characteristics of QoS, these characteristics form three logic planes (Fig. 3.1). There are a control plane, a data (information) plane, and a management plane.

General approaches (methods, techniques) used for providing required QoS for these planes are shown in Fig. 3.2. Let us discuss these approaches more thoroughly.

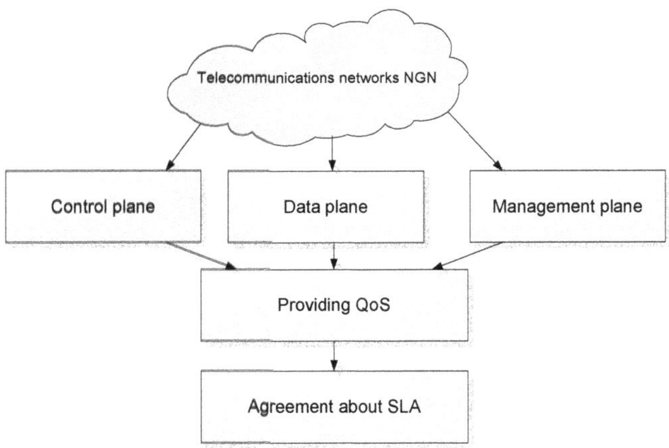

Fig. 3.1 Logic structure for formation of SLA

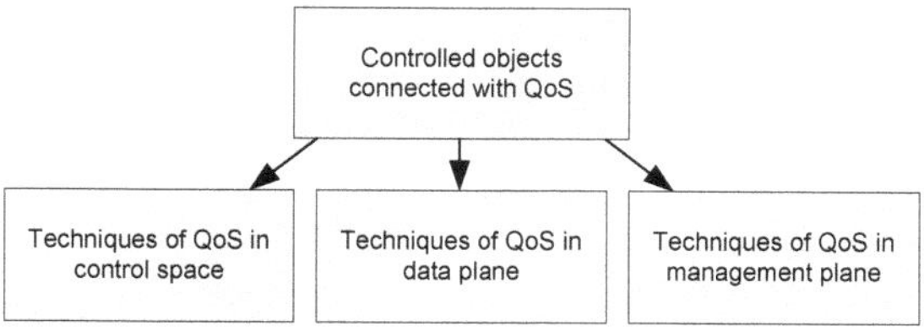

Fig. 3.2 List of control objects connected with techniques of QoS

Techniques for the control plane. These techniques are connected with the path of transmission of user's traffic. There are three main techniques, discussed below.

The technique CAC (Call Admission Control) controls new requests for traffic transmission through the network. It determines either this traffic results in the network congestion or in degradation of the existed level of service for already existed traffic.

The technique of QoS routing provides the choice of a route satisfied to a required quality of service for a given data flow. As a rule, only one (or maximum, two) network characteristic is taken into account to make this choice. These characteristics are the performance and delay, or price and performance, or price and delay and so on.

The third technique is called the resource reservation. In IP–oriented networks, the most typical resource reservation technique is based on the protocol RSVP. This protocol provides the possibility for implementation of the route before the beginning of the session.

Techniques for data plane. This group of techniques deals with user traffics. It includes such techniques as the buffer management, congestion avoidance, packet marking, queuing and scheduling, traffic classification and shaping. Let us discuss them in details.

The technique of buffer management is reduced to the control of packages which are waiting in a transmission queue. The most important goal here consists in minimizing the average length of a queue with providing a high rate for channel utilization. The second goal is providing of the fair distribution of the buffer space among different data flows. Nowadays, the techniques for active buffer management are wide spread. A typical example of this approach is the algorithm of probabilistic preliminary revealing of possible congestion. This algorithm is named RED (Random Early Detection). When this algorithm is used, the packages are cutting from the queue on the base of its average length. The longer is a queue, the higher is a probability for refuse of service for a given package.

The mechanism of congestion avoidance support the level of network's loading a bit below of its throughput. The usual way for avoidance of congestion is reduced to diminishing of the network traffic. As a rule, an instruction for traffic decrease is oriented on cutting the sources with low priority. One of the examples is the window mechanism used in the TCP (transmission control protocol), which is typical in the Internet.

The approach of packet marking is used to denote a particular service level for different packages. As a rule, the marking is made into an input communication unit (input hub). To do it, some value is introduced into a special header field.

The main goal of techniques from the group of queuing and scheduling is the choice of a package for transmission from the buffer into a communication channel. The majority of service procedures (disciplines, or schedulers) is based on the rule FIFO (first in – first out). Some other approaches were proposed to provide more flexible service disciplines. As a rule, they are based on existence of more than one queue. It is worth mentioning the discipline of privileged (high-priory) service. Another example is a mechanism of weighted fair queuing (WFQ) when the restrictive throughput of a hub is distributed among some data flows (queues) in respect with requirements of each queue. There is such a discipline which is based on classification of data flows according with their classes of services. It is named CBQ (Class-Based Queuing) discipline. The data flows are classified and placed in the different queues of a buffer. Each queue disposes some part of throughput depending on its class. The queues are served using a cyclic approach.

The techniques of the traffic classification are connected with classification of packages on the input of a network. Classification is executed using common requirements to their service. The classification is done on the access-node (or boundary router) of the network. It allows finding packages placed into the same data flow. Next, the traffic is processed using a normalization procedure. The special mechanism (Traffic Conditioning) is responsible for this normalization. The traffic normalization supposes the measurement of traffic parameters and comparison of measurement results with required values stated in SLA. If the conditions of agreement are violated, some part of packages can be casted away.

The goal of traffic shaping is control of speed and size of data flows from the input of a network. The initial traffic is put through special formatting buffers. It allows more predictability of traffic characteristics. Two main mechanisms are known to solve this problem. The first of them is named "a leaky bucket" and second has name "a token bucket". The "Leaky Bucket" algorithm executes regulation of the speed for packages going from a hub. The output performance of the hub is a constant value which does not depend on the speed of the input flow. When the bucket (buffer) is overflowed, all excess packages are thrown away.

As opposed to this approach, the "Token Bucket" algorithm does not regulate output speed of a hub, as well as does not throw away excess packages. The input and output speeds can be equal, if there are tokens into the corresponding buffer. The tokens are generated with some definite speed and they are accumulated on the bottom of the "bucket". There are two parameters characterizing this approach. They are the speed of generation of tokens and the size of the bucket. The packages cannot leave the hub, if there are no tokens in the bucket. Otherwise, a pack of packages can left a hub after elimination of some amount of tokens.

The management plane of QoS includes techniques responsible for maintenance, administration, and control of a network in respect to user traffic. It includes some specific techniques. One of them is metering, which provides the checking parameters of traffic service. For example, such parameters can be measured as the real speed of the data flow. Next the results of measurement are compared with required values stated by the SLA. After analysis of these results, some specific procedures can be applied, for example, either Leaky Bucket or Token Bucket mechanisms.

The central place into implementation of these mechanisms takes the signalization protocol RSVP (Resource reservation Protocol). This protocol provides reservation and management of resources.

To provide the required service quality on the stage of carrying of packages, the RSVP protocol should be supplemented by one from existed routing protocols, as well as some set of mechanisms of traffic control. These mechanisms include access control, traffic classification, control and planning of queues, and so on.

There are three models of service providing, namely the best, integrated, and differential. We can denote them as Best Effort, IntServ, and DiffServ Service, respectively. Some peculiarities of these models are shown in Fig. 3.3. Let us discuss them in details.

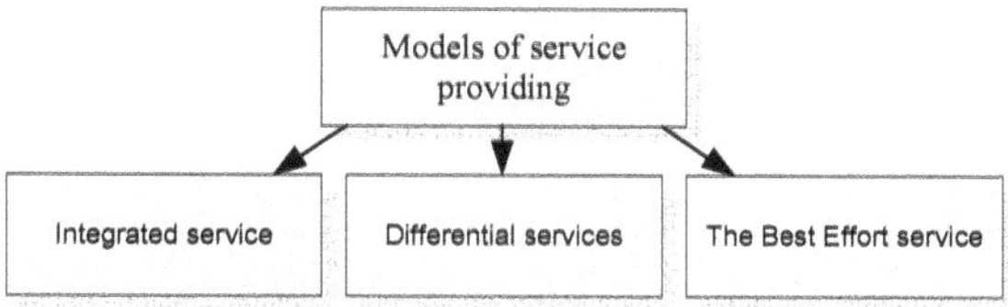

Fig. 3.3 Basic components of models for service providing

The models of Integrated Services (IntServ) have the following features. They use integrated reservation of resources before starting transmission for a given data flow. They possess such a negative feature as the redundant reservation of the channel capacity for some flows. It has the negative influence on the QoS for other flows, even in periods when this reserved capacity is not necessary. They give strict guarantees for the service quality and the low level of scalability. They have no special tools for providing QoS for some macro-flows; it restricts significantly the area of applying for IntServ. This model is connected with the growth of loading on routers to give a possibility for operation of the IntServ service (especially in high-speed highway networks). The use of this service requires tremendous alterations in the router software for recognition of network applications (applets).

The models of Differentiated Services (DiffServ) are oriented on differential service providing for some assemble of traffic classes. It provides the regulation of QoS only for macro-flows. Because of it, these models are oriented on relatively small amount of service classes. They are characterized by the lack of guaranteed realization of the promised requirements for macro-flows.

In the case of Best Effort Service, the fair distribution of resources is guaranteed. But this model does not support the mechanisms of control for the network resources and their distribution. Besides, it allows occurrence of congestion if there are sharp traffic hits, as well as does not guarantee the data delivering. But if there is no need in the real time data delivering, than this model is very efficient.

3.2 Peculiarities of RACS and RASF Approaches

The Resource and Admission Control Sub-System (RACS) executes the control functions for access network and boundary node of the kernel execution level. The kernel execution level is a part of the network with routing mode using IP protocol. The access network is a part of the network where traffic is aggregated or spread without use of the dynamic routing. The control of resources inside the access network belongs to the second level of the model of open system interconnection (OSI). In the RACS conception, the control of resources is not considered for the level of network's kernel.

The direct control of resources is directly executed by some network element determined the traffic for the second level, as well as by the network element placed on the boundary of the transport network. The component A-RACF (Access Resource and Admission Control Function) executes some functions connected with the access to network resources, as well as control functions of these resources. The component SPDF (Service-Based Policy Decision Function) executes control using the access politics to service on the boundary of the level of the network kernel. The RACS architecture (Fig. 2.4) does not receive any information about the topology of the kernel network. The management of QoS is made in the mode "Push" (using the Ponselle principle). In this case, the controlling component (either A-RACF or SPDF) sends instructions to the transport equipment.

There are no serious contradictions between the conceptions RACF (Resource and Admission Control Function), proposed by the International Telecommunications Union (ITU), and the architecture RACS, proposed by the European Telecommunications Standard Institute (ETSI). It can be explained by the fact that both institutions, ITU and ETSI, were in close cooperation during development of corresponding standards (architectures). But some differences can be found.

One of them is a part of the network where the control takes place. In contrast to RACS, the architecture RACF considers the management process for all parts of the network to provide a required QoS. Besides, the architecture RACF provides more scenarios for management of the network resources, than it is in the RACS architecture. Because of it, the RACS architecture should be treated as a part of the RACF architecture.

The ITU determines the management architecture of QoS in its standard. The main idea of the management architecture of QoS is independence of the transport level from the service level. For example, when the speech is transmitted using the IP protocol through the Internet the speech traffic is transmitted after the procedure of signal information exchange after fixation the connection between a workstation and signalization server. Besides, this traffic can run further through a network of a mobile operator. But the operator cannot transfer the given traffic with the highest priority and get the additional gain, because now there is no mechanism for request and guaranteed providing the required QoS on the network level.

To solve this problem, the ITU proposed to distinguish the levels of transport and services, given them inter-independence. In accordance with the conception of independence for these levels, the required network resources are provided by the

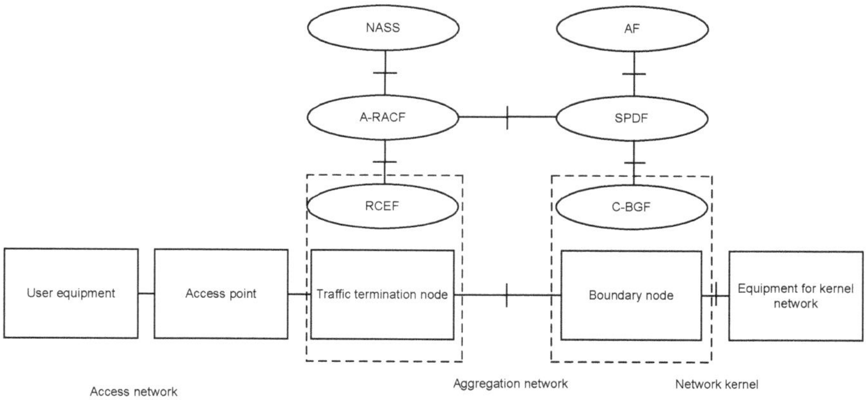

Fig. 3.4 Architecture of RACS subsystem

network after obtaining the instruction from the service level. In turns, the service level takes responsibility for the exchange of signaling messages among the applications. The transport level is responsible for both reliable transmission of the data packages and traffic's control. The service level can be represented either by some application server or by a system.

The control function of the transport level serves for interconnection between the levels of services and transport. This function permits the service providing based on analysis of the state of network resources and access politics, which are established by the operator to a given user. This function controls the network equipment to provide the required services. The function of RACF determines the availability of network resources and executes their control. The RACF architecture is shown in Fig. 3.5.

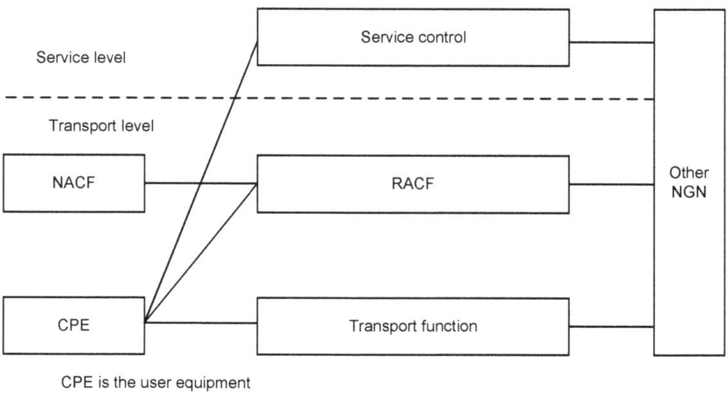

Fig. 3.5 Architecture of RACFsubsystem

The service control function (SCF) is responsible for transmission of signal data during the time of setting for a given communication session. This function requests the required level of QoS in given element of RACF. This element determines the availability of network resources for providing the required QoS and, next, controls the network equipment.

The function of control for network addition (network addition control function, NACF) supports the profile with given level of QoS for a given user. During the procedure of call identification, this function checks availability of needed access network resources. The functional architecture was designed taking into account the principle of independence from the user site. Because of it the RACF architecture can be implemented into the access network, as well as in the network kernel level.

The element RACF includes two functional blocks. Fist of them is responsible for execution of required rules and politics (PD-FE), whereas the second controls the resources of the transport level (TRC-FE).

The PD-FE block determines the ability for given service providing. To do it, the following data are checked: the user profile in the access network, the service level agreement, politics, priorities, existence of required network resources. When the request is received by the PD-EE element, it sends information for transfer of the traffic to the network equipment to provide the required portion of network resources. This information includes the following items: the instructions for the gateway control with the order about either permission or barring for transfer of these data; marking (labelling) of packages of the data flow; the data about IP-addresses and ports for execution the NAT functions; control of the transmission speeds; the mode of operation for a interwork screen (shield) for the filtering of traffic; the order of data transmission (the choice of a route and a network for service providing with the guaranteed quality).

The functional element for decision making, PD-FE controls the equipment of the access network using the functional element for providing guarantee of appointed rules, PE-FE. The element PE-FE is placed on the boundary of a regional network. In the real communication network the functions of PE-FE element can be executed by the following equipment: the boundary gateway for session control (SBC, session boundary control); the cable modem termination system (CMTS); the boundary router.

Therefore, the functional element of decision making, PD-FE controls the QoS of a network using the PE-FE element, placed on the network boundary. The functional element for control the resources of the transport protocol, TRC-FE tracks the state of network resources in the regional network. It makes a decision about the service providing on the base of the data about the availability of network resources.

The functional element TRC-FE targets the control of access network's resources basing on a transport protocol in use. The block PD-FE controls the resources of the transport network without dependence on the transport protocol. It is worth pointing, that there are no specific mechanisms for implementation of the functions of the element TRC-FE in the current version of ITU standard.

Nowadays, the data transmission technology is used which is based on the stack of TCP/IP protocols. It uses the approach of packet switches and combines many different control mechanisms. From the system theory's point of view, the implementation for these mechanisms can be divided by two approaches. First of them is connected with control of the structure of TCS, the second is connected with the functional control.

3.3 Control of the Structure of Telecommunication System

It is well-known that the structural properties of systems are reflected in the peculiarities of a network. These peculiarities include the systems of communication centrals and lines (links) used for their connection. Remind the definition of the system: a system is a set of interconnected elements. The network is a backbone object. It means that there is no system without this network.

There are some levels of definition of the structure of network. It could be defined on the physical level (communication centrals and links), or on the channel level (multiplexing), or on the network level (routing). Obviously, each of these levels gives a different structure of a network. It gives the different content for control on these levels. This structure can be either static or dynamic, changing under the influence of the traffic and corresponding control protocols.

Very often, the physical structure of a network remains constant in time. But the deficit of different physical resources (such as the frequency range, time, and spatial-polarized resources) requires the applying dynamic distribution of these resources. It concerns such resources as the spatial-temporal access, spatial-temporal encoding, repetitive usage of frequencies, adaptive antenna arrays and so on.

The network structure for the channel level can be either static or dynamic. The necessity of transfer to dynamic structure arises under the traffic termination.

The network structure on the network level is rather dynamic, for both dynamic and static routing and locking (use of getaways). The static routing is based on the principle of minimum amount of nodes per a route (the principles of the shortest path). The dynamic routing is connected with the choice of the best path in accordance with current data about the loading of a network. In both cases, the network structure is changed by its traffic (Fig. 3.6).

Thus, the modern telecommunication network should be treated as a network with dynamic changeable structure. Different control procedures are used for providing such a dynamic structure. The necessity for usage the sequences of different procedures is connected with various alterations made for the data during their movement from a source to a receiver. This movement passes through different parts of a network having different control tasks.

The main methods used for implementation of control procedures for a network structure are software methods having the threshold nature. The essence of this nature is in the changing control structure after achievement some appointed threshold. The main role here the packet headers play, where the addresses of sender and receiver are pointed, as well as some other data permitting execution some functions during the packet's progress in its route. In the same time, the optimal formalized methods are used more and more often. These methods provide maximum improvement or optimal choice of the network structure using some criteria.

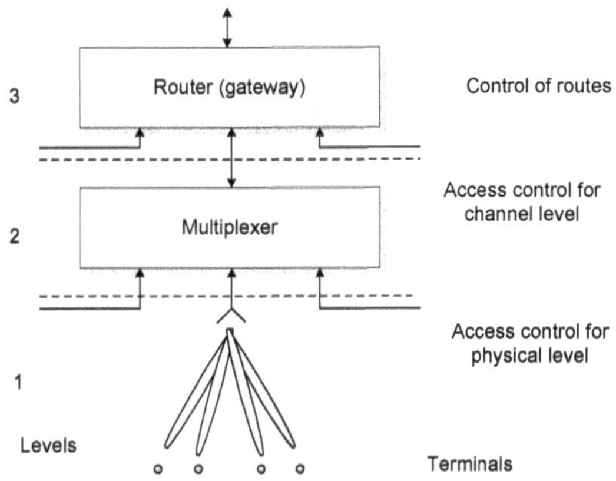

Fig. 3.6 Logic of distribution for control of network structure

Optimization of network structure. The necessity in optimizing the control of a network structure arises when the usual software-based control methods do not provide required level of QoS. The most critical (bottleneck) network element is the router, where tremendous delays are possible, as well as other losses. The task of exhaustive search leads to finding the best path, but this task is NP-complete and cannot be used due the lack of time in a real network. Because of it, some optimal procedures are developed for finding the optimal routes. One of them is the OSPF protocol, where OSPF stands for open shortest path first. This protocol uses an optimization tasks belonging to the class of tasks with distribution of resources. It is the task of linear programming.

The search has a one-step nature. It results in the necessity of the problem solving for each new step. There are known some attempts to solve this task as a task of the dynamic programming. But is requires the tremendous size of housekeeping data; it makes the reasonability of this approach use very questionable.

Optimization of dynamic procedure for control of TCS structure. Let us use more constructive method taking into account some constraints for the level of QoS. To avoid the NP-completeness of the task of redistribution of traffic, let us use a recursive algorithm for minimizing congestions in the network under fulfilment of some conditions.

The main idea of the algorithm is the following one. It is proposed to use the centralized control strategy for the totality of a network (or for its part). It is shown in Fig. 3.7.

First of all, the reduced values of load are found to construct the matrix with a zero diagonal (3.1).

$$\|\hat{T}_{ij}\| = \begin{bmatrix} 0 & \hat{T}_{12} & \hat{T}_{13} & \hat{T}_{14} \\ \hat{T}_{21} & 0 & \hat{T}_{23} & \hat{T}_{24} \\ \hat{T}_{31} & \hat{T}_{32} & 0 & \hat{T}_{34} \\ \hat{T}_{41} & \hat{T}_{42} & \hat{T}_{43} & 0 \end{bmatrix} \tag{3.1}$$

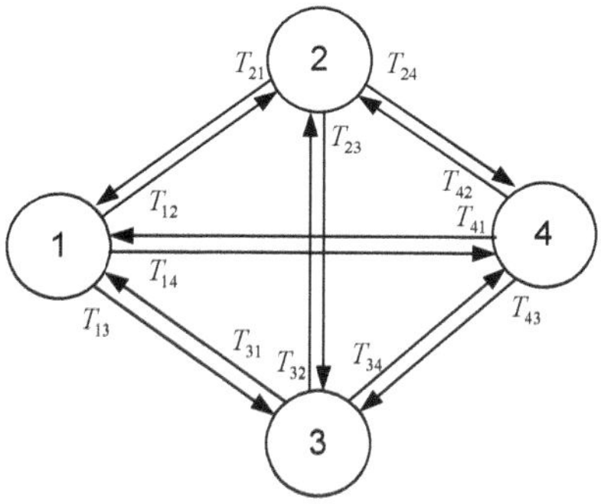

Fig. 3.7 Construction of matrix of reduced traffic for centralized control

Nondiagonal elements correspond to reduced load (traffic) for corresponding direction of communication in the network, $\hat{T}_{ij}$. It is determined by the following equation:

$$\hat{T}_{ij} = \frac{T_{ij}(t)}{v_{ij}}. \tag{3.2}$$

In 3.2 the symbol T_{ij} stands for the corresponding load in the instance *t*, the symbol v_{ij} means the throughput of communication channel *ij*. Each following time interval is connected with redistribution of the load from the most loaded channels *ij* on the unloaded by-passes including two sections, *ik* and *kj*. The amount of these sections can be more than two.

The proposed algorithm can be used either in the case fully connected network, or for the network with arbitrary connections. The restrictions of connectivity are not principal and have no influence on the operation of the algorithm. Let us point out that for MPLS-TE technology the full connectivity of the network can be reached on the base of logic methods. It means that some tunnels (virtual channels) can be created among all inputs and outputs of distributed elements. It results in the full connectivity. The main advantage of this method is its ability in preventing the congestion of a network, as well as its ability for restart after failures, collapses of some network elements and directions.

Let us discuss a proposed algorithm for directed search of overloaded paths and redistribution their loading on other bypass channels. Let us find a solution for the algorithm of centralized control for finding adjusting values for loading existed for each path. These values are used to construct the matrix of adjusting loadings (3.1), where nondiagonal elements show adjusting values for network loadings. The control task is reduced to finding appropriate values for these nondiagonal elements.

Let v_{ij} be a throughput of a communications channels between the nodes *i* and *j*, and let $T_{ij}(t)$be a corresponding loading in the instant of time *t* . If the loading between two channels is too big, it is necessary to use some bypasses. Let us call relation (3.3) the adjusting loading of the channel *ij*, where this relation is the following one:

$$\hat{T}_{ij} = \frac{T_{ij}(t)}{v_{ij}}. \tag{3.3}$$

Obviously, that the channel is completely loaded when there is $\hat{T}_{ij} \to 1$. In this case a necessity arises for rerouting some part of the loading, using bypass routes.

Let F_{ij} be the given data flow from the node i into the node j. If there is a direct connection between the nodes i and j, then the adjusting data flow from the node i into the node j is determined by the following relation

$$\hat{F}_{ij}^{i}(t) = \frac{F_{ij}(t)}{v_{ij}}. \tag{3.4}$$

In the case when the path from the node i into the node j runs through some intermediate channels, then the adjusting data flow for each channel is determined by the following expressions

$$\hat{F}_{ii+1}^{ij} = \frac{F_{ij}(t)}{v_{ii+1}};\ldots;\hat{F}_{kj-1}^{ij} = \frac{F_{ij}(t)}{v_{kj-1}};\hat{F}_{j-1j}^{ij} = \frac{F_{ij}(t)}{v_{j-1j}}. \tag{3.5}$$

Thus, the loading for each direction $\hat{T}_{ij}$ consists from the loading, transmitted between the nodes i and j, that is the value $\hat{T}_{ij}^{i}(t)$ and the totality of loadings created by rerouted traffic. So the following equation can be written:

$$\hat{T}_{ij} = \hat{T}_{ij}^{i}(t) + \sum_{k=1}^{n-1} \hat{F}_{kj}^{ij}. \tag{3.6}$$

In (3.6) the symbol n stands for the number of nodes in the network, $k \neq i$.

To control the distribution of control flows, it is proposed the following spatial-temporal recursive procedure. It is executed for each of k steps starting from the node i having the maximum loading and moving to the node j. Obviously, the adjusting flow can be represented for each instant of time k by the following expression:

$$\hat{F}_{ij}^{i}(k) = \hat{F}_{ij}^{i*}(k) - \Delta\hat{F}_{ij}^{i}(k). \tag{3.7}$$

The value $\Delta\hat{F}_{ij}^{i}(k)$for the step k shows for each iterative step which part of loading will be redistributed from the most loaded channel to the bypass channel having two paths. This value can be set up either taking into account the adopted control policy or using some empiric rules. It should provide the stability of operation for the given time interval under the increased loading. The practice shows that the reasonable value of this parameter is equal to 20%. It means that remaining loading for this direction can be found as

$$\frac{\hat{F}_{ij}^{i}(k) - \Delta\hat{F}_{ij}^{i}(k)}{\hat{F}_{ij}^{i}(k)} = 0.8. \tag{3.8}$$

The following step of redistribution is connected with account of values for throughputs v_{ij}. The values should be analysed for determining their relations for the first and further steps of digitization.

The time step number $k+1$ is connected with transition of the procedure on the following step, taking into account the value of adjusting flow from the previous step. Acting in the same manner, we can find the following recursive procedure for estimation of a state:

$$\Delta \hat{F}^{i}_{ij}(k+1) = \Delta \hat{F}^{i}_{ij}(k) + \mu \left[\hat{F}^{i}_{ij}(k+1) - \hat{F}^{i*}_{ij}(k)\right]. \tag{3.9}$$

The value μ is chosen from the conditions of stability for the given procedure. The value μ is connected with the convergence rate of the recursive procedures (3.8) or (3.9). The value of difference in square brackets is named the residual. The less is the correlation interval τ_{cor} between the random values $\left[\hat{F}^{i}_{ij}(k+1) - \hat{F}^{i*}_{ij}(k)\right]$, the closer to 1 the value μ should be. On the other hand, decrease of correlation for sections on the steps k and $k+1$ leads to corresponding decrease for the value μ. As experience shows, these sections should be less than one tenth of correlation interval (that is $\Delta t \leq 0.1\tau_{cor}$).

One of the features of discussed procedure is the necessity in accounting current loading for all parts of a network. It can be easy done for the network with coordinated control. But some technologies make impossible (or very difficult) the organization of centralized control. Because of it, the proposed procedure can be modified for local control for each network device. In this case, each specific router can have only data about the loading for channels directly connected with this very router. In this case the loading for one from two paths of a bypass is taken into account. It can lead to nonoptimal redistribution. But the discussed method possesses one positive property: there is no need in usage additional loading because of the absence of the official traffic.

3.4 Control of Functional States of TCS

The functional states of telecommunication systems are reflected by current modes of network elements creating the given network. These state, $x(t)$ can be changed in time under influence of some factors. The desired states can be set up using the control procedures either for influences or states.

The control function for a given object determines a control mechanism for this object. It includes the state monitoring, control of state changing, instructions for state changing control. Thus, it is enough to take into account only a current state $x(t)$ and the rate of change for the given state, $dx(t)/dt$.

The control of the object state determines standards for representation of a current state of an object. This state can be changed under the influence of control actions. There are three functional areas of control.

First of them is the monitoring of operability for a given object. It is determined by lack or existence of required resources, which can be used by a control system. Very often, only two states of an object are enough, namely permission (enabled) and barring (disabled) of resources control. In common case, it is possible to use either conceptual or discrete modes for resources control.

The second functional area is the usage or loading of a unit. It determines whether the unit has some loading. It gives information about availability of free resources. Three states of an object are possible here, namely: the object is free from loading (idle), the object has some loading (active), or it is intensively used (busy);

Such area, as the administrative (management) state describes possibility for usage these or those resources. This state is divided by three phases: the access to resources is blocked (locked), the mode of turning off or stopping (shutting down). Even if the resources are blocked, it could be the possibility for their control. For example, the state of administrative blocking appears if an incorrect user password is entered to the access to the control system.

Each of discussed states of controlled objects has different characteristics described by some attributes. The attributes used for characteristic of operability and usage of an object should be easy readable for user of control systems. It concerns the systems with situational control methods, too. The attribute for administrative state should be accessible from the side of a control system.

The different approaches for interrelations between different objects are regulated by the special relationship management function. It is determined by ITU (Recommendation .732). Using this function, it could be determined which object sends the control instructions and which object receives and executed these instructions.

3.5 Basic Network Control Protocols

If the control is implemented as an automatic (self-acting) procedure, then some mathematic model should be constructed for the state $x(t)$. The model should reflect the dynamic properties of this state such as $(dx(t)/dt)$, the optimality criterion for state control, $J(t,x,u)$, and restrictions (constraints) for the used variables $x \in X$, $u \in U$, $t \in T$. It leads to the standard optimization procedure.

The control procedure can be either situational or formalized. But in both cases the decision about any control action is made on the base of monitoring of the current state of a controlled object. Some methods are developed for providing such a monitoring.

For example, it can be used the Common Management Information Protocol (CMIP) determined by the standard ISO 9696. As it is known, ISO means the International Organization for Standardization. This protocol provides interrelation for open system on the application layer. The procedure Agent/Manager is implemented in this protocol. The program-manager side issues control instructions, whereas the protocol machine receives requests for service providing in the control area. The side of the program-agent is provided by all necessary data. But this protocol turns out to be very cumbersome and difficult for implementation.

Next is the protocol SNMP (Simple Network Management Protocol), which is implemented in three versions. They are the versions SNMP-1, SNMP-2, and SNMP-3. It was spreading from 1993 as a method for management of TCP/IP network. The structure of interrelations for the SNMP is shown in Fig. 3.8.

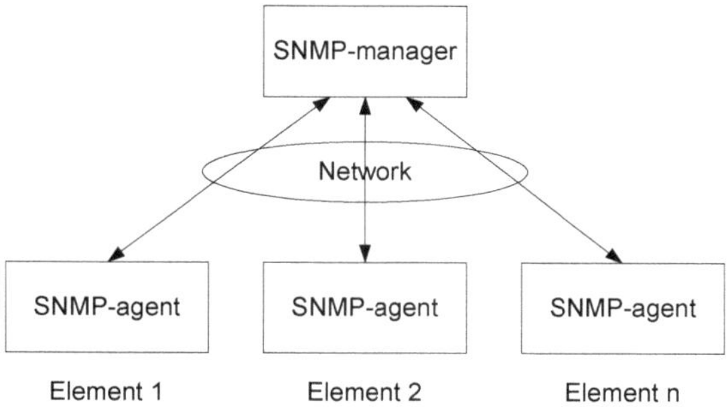

Fig. 3.8 Organization of control for NMS System

This protocol has the following properties listed below. It determines a network as a collection of network control stations and network elements such as main machine, gateways, routers, terminal servers. These components provide administrative connections among control stations and network agents. The SNMP belongs to protocols of application level, it provides data exchange among network elements. It gives some important data to make some decisions. The network administrator (manager) has a user program providing virtual connections with programs of network agents. They cooperate in the mode "request - replay" using the principle of polling. A SNMP-agent occupies some remote network device and gives some data about the state. The SNMP-agents can be set up on any network device (server, printer, router, hub, and so on). These agents organize the virtual data array named as MIB (Management Information Base).

The programs-managers execute the following functions: gathering of information about states and failures, checking of trend for states of network elements and devices, interruptions during the data receiving, filtration of inessential interruptions if a failure turns up, tools for control of reconfiguration of a network, and MIB-compiler for adding information about new elements.

The second protocol is RMON (Remote Network Management Protocol Information Base). It can be viewed as the extension of the SNMP protocol providing also data gathering and analysis for states of the objects of monitoring. Additionally to the functions of SNMP, the RMON protocol not only registers the evens for a particular object with some agent, but characterizes data about the traffic among the network devices. This protocol allows gathering data about operation of the network and setting warnings about undesirable situations.

There are two different versions of the RMON. The version RMON-1 operates on the medium-access control (MAC) level. The protocol RMON-2 operates on the network level and on higher levels. The specific feature of the version RMON-2 is its ability to identify either sender or receiver from the opposite site of a router. It provides gathering statistics for all hosts to know who requires the access for a given part of a network. The places for operation of both protocols (RMON-1 and RMON-2) are shown in Fig. 3.9. In this figure the symbol BOC stands for the Bell Operating Company.

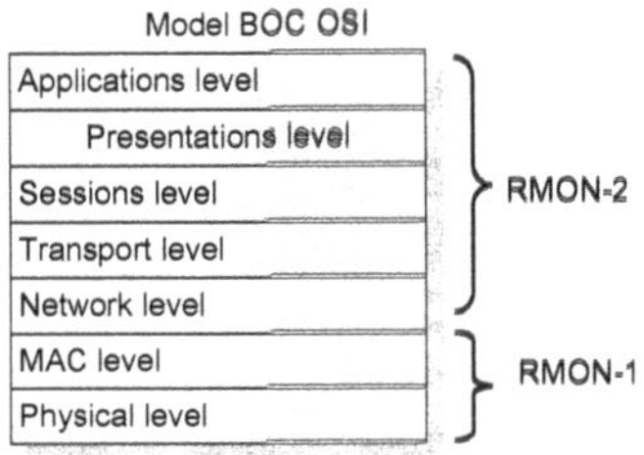

Fig. 3.9 Distribution of functions for RMON-1 and RMON-2 protocols

The protocols RMON-1 and RMON-2 cannot be viewed as subsets or replacements for each other, they merely supplement each other.

Both protocols SNMP and RMON have the following positive features:

1. Simplicity, accessibility, independence from manufactures. They have small set of instructions which can be implemented very easy. Because they require minimum resources, they can be used for the networks of personal computers.
2. They require minimum resources for organization of monitoring. Because of it, it is possible to use personal computers as servers.

The third protocol is the Net Flow (Cisco Net Flow) developed by the Cisco corporation. It operates on the base of Agent/Manager technology and it is used for centralized data gathering about the network traffic. It can produce bills on the base of the real usage of network resources. Also it provides safety of the network and can execute monitoring of attacks, as well as the general network monitoring.

Architecture of the Net Flow protocol includes three main components. The first of them is a sensor (or data exporter) named NFE (Net Flow Exporter). It provides data gathering about sessions and sending it to a collector. The second element is the collector NFC (Net Flow Collector). Last element is data processor named NDA (Net Flow Data Analyzer).

The Net Flow protocol possesses some positive features in organization of monitoring. Apart from such functions as data registration, aggregation and tariffing, the agent for this protocol can check the access of prescribers into IP network. It is possible to terminate the services for given prescribes if there are no money on their accounts. The functions of on/off are implemented by external procedures, which are controlled by the Net Flow agent's access control system (ACS). It permits to reach the maximum level of flexibility to integrate the agent into existed access control systems.

As well as the Ethernet-agent, the Net Flow agent can operate in the SAFE mode, when a channel between the server and the agent either is not safe or has no required throughput. In this case a local access server (where the agent operates) executes primary registration and storing of required data. Such an approach minimizes the amount of transmitted data between the server and agent. Also it permits interception of access control for prescribers when there is no connection with the central data storehouse (data base). It permits blocking and unblocking of prescribers using local data known in the instant when the connection with the central data base was broken down. After recovery of this connection, the automatic replication is done for data bases of both agent and server.

This approach has some drawbacks, too. First of all, operation with use of user datagram protocol (UDP) can result in loss of messages of the trap type from agents to managers. In turn, it can lead to absence of response from the control site and, therefore, the control quality will be diminished. If an attempt is made to use the transport protocol in this case, it can lead to communication loss with huge amount of embedded SNMP agents. Next, the considerable part of the channel bandwidth is occupied by data transmission between agents and servers. Next, to organize wide and complex monitoring in some big network it is necessary to buy very expensive software and hardware. At last, this protocol makes very high demands to performance of servers.

Recommended Literature

1. CISCO. Cisco Networking Essentials, vol. 2. Cisco Systems, Curriculum Development, Team Worldwide Education (1998)
2. CISCO. Cisco Wireless Control System Configuration Guide. Cisco Systems (2007)
3. Conti, M.: Wireless Communications and Pervasive Technologies. John Willey and Sons (2005)
4. IEEE. Telecommunications and Information Exchange between Systems. Local and Metropolitan area networks –Specific requirements Part 15.4: Wireless Medium Access Control (MAC) and Physical Layer (PHY) Specification for Low-Rate Wireless Personal Area Networks (WPANs). IEEE Standard for Information Technology (2006)
5. Rakley, S.: Wireless Networking Technologies. Newnes/ Elsevier, Amsterdam (2007)
6. Tanenbaum, A.: Computer Networks. Prentice Hall PTR, Englewood Cliffs (2002)

Chapter 4
Mathematical Models and Principles of Control in TCS

Abstract. This chapter is devoted to mathematical models and control principles used in telecommunication systems. Particularly, there are discussed static and dynamic, deterministic and stochastic models of controlled systems and decision-making techniques of control modes. The methods of state variables are used to construct these models. Some criteria of control optimality are discussed, such as the compatibility criterion, leading to the principles of guaranteed quality, different criteria of optimality and preference criterion. The criterion of minimum of square deviation is discussed thoroughly. Both the algorithm of optimal control and main principles of control systems' construction are considered. These principles are based on methods of Ponselle and Watt. The decomposition theorem is formulated.

4.1 Procedures for Control Decision Making

The majority of known control methods used in technique has the casual nature. It means that any action is executed as an answer for some cause (or motivation). It could be not true for social systems. For example, making the assurance policy is an answer on a disaster, but the disaster does not take place yet. There are similar approaches in technique too, when so called invariance decisions do not obey the principle of causality. In this case, some redundant resource is included into a system (or in signals). It allows fulfilment of the required system functions operating in unwanted situations. For example, the signals with redundant base such as CDMA possess the invariance property of insensibility to the class of effecting concentrated interferences (there are no interferences, but the reaction already exists).

To have a reason for starting control procedures, some corresponding information is necessary. After analysis of this information, some control decision can be taken. There are different data which can be treated as this necessary information. First, it could be data about either a structure of a state of a system $S(x,t)$, allowing making decision about changing either the state or the structure of the system. These changing should satisfy some criterion of efficiency, $J(x,t)$. Next, it could be

V. Popovskij et al.: Control and Adaptation in Telecommunication Systems, LNEE 94, pp. 37–53, 2011.
springerlink.com

data about existence and/or parameters of external influences, $y(t)$, which make impossible or difficult the proper system operation. At last, it could be some schedule (cyclogram, timing regulation) required execution of some actions in some instance of time (turning on, turning off, change of mode, and so on).

Thus, it is necessary to have some corresponding data to make control decisions. These data can be obtained as results of monitoring (observation). The objects of monitoring can be either the system state $x(t)$ or state of influences. As it was pointed out, the data can be represented by some schedule. Let us discuss the procedures of decision making based on monitoring results in details.

The necessary data are taken from some measuring devices. For example, if the SNMP protocol is used, the data are obtained from agents making monitoring for these or those network elements. The watch equation represents the value of observed signal $x(t)$ against the background of the noise $v(t)$:

$$y(t) = x(t) + v(t). \tag{4.1}$$

Let us think that the strategy of system operation is reduced to analysis of the signal $x(t)$ and generating control influence $u(t)$. This task of signal detection is reduced to the checking of the following statistical hypotheses:

$$\begin{cases} H_0 : y(t) = v(t), & \text{if the signal does not exist,} \\ H_1 : y(t) = x(t) + v(t), & \text{if the signal exist.} \end{cases} \tag{4.2}$$

Obviously, some threshold h should be chosen on the base of some used criterion. If this threshold is exceeded, then the hypothesis H_1 is taken, otherwise the hypothesis H_0 is true. In other words, the observing the random process $y(t)$ should result in making deterministic decision (upper or below the threshold, $y(t)$).

Due to the random nature of both the signal $x(t)$ and the noise $v(t)$, the errors are possible under the choice of the threshold h. There are two kinds of such mistakes. The error of the first kind is acceptance the hypothesis H_1, whereas the hypothesis H_0 should be accepted (it is a false alarm (FA). The error of the second kind is acceptance the hypothesis H_0, whereas the hypothesis H_1 should be accepted (it is a missing of object (MO). The densities of distribution $P_1(x)$ and $P_0(x)$ are shown in Fig. 4.1. These densities corresponds to hypothesises H_1 and H_0. In this case the choice of threshold h corresponds to the following choice criterion: the both probabilities of the false alarm P_{FA} and missing of the object P_{MO} have minimum possible values and are equal. Such a criterion is called "the criterion of an ideal observer".

These probabilities are determined by the areas under the curves of densities $P_1(x)$ and $P_0(x)$ to the left and to the right from the threshold level:

$$P_{FA} = \int_h^{\infty} P_0(x)dx, P_{MO} = \int_{\infty}^{h} P_1(x)dx. \tag{4.3}$$

The probability of the general error is equal to $P_{ER} = P_{MO} + P_{FA}$. When it is necessary to minimize one of probabilities (4.3), the value of threshold h is shifted

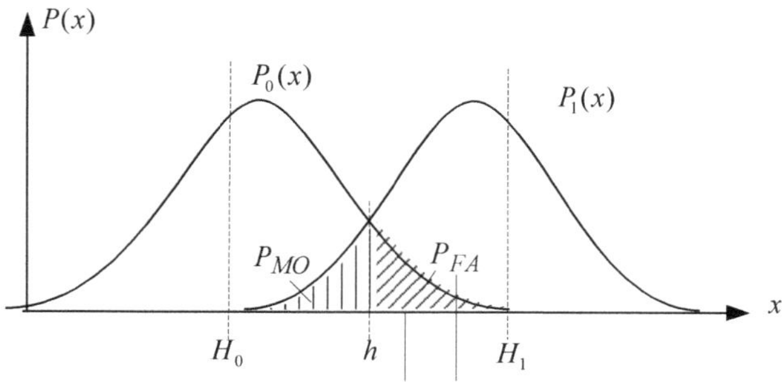

Fig. 4.1 Standard densities of probabilities $P_1(x)$ and $P_0(x)$ and errors

either to the right or to the left, respectively. In the same time, if one of the probabilities is minimized, then the second one is increased together with the general probability P_{ER}.

The task (4.2) can be generalized for the multi-alternative case when a lot of hypothesises are considered: $H_1, H_2, \ldots, H_n$. It leads to increase for the number of thresholds (now there are thresholds $h_1, h_2, \ldots, h_n$). Passing to such an approach for multi-alternative decision making, it is possible to choose a necessary control procedure for a given hypothesis H_i.

Besides of the discussed criteria method of decision making, they use the situational approach. It differs from the criteria method, because situations for decision making are analysed instead of hypothesises (4.2). The possible decisions are formed as a function depended on situations and kept in the choice catalogue. The rules of situational choice are based on logic procedures of the type "if-then". The decisions can be found in the knowledge base (if all possible situations are known). Obviously, the efficiency of situational decisions is restricted, because it is a priori impossible to know all possible situations and place them into a catalogue. But the final decision here belongs to PMD. It means that the defining role in this case belongs to experience and engineering intuition of PMD.

4.2 Types of Structures of Controlled Systems

There are two comprehensive characteristics of any system, even so complex and dynamic as TCS, namely the structure and function. Very often, the term "system" is replaced by the term "network". In this case it is supposed that the network is an adequate model of the particular structure and represents its backbone properties. As it is known, the term "backbone properties" means "properties created a system".

The system structure, as a rule, is represented either by some graph, or by matrices (the incidence matrix, the connectivity matrix, the adjacency matrix, and so on), or in the form of set-theory description, or with help of topological methods, and so on.

There are some standard structures used in telecommunication systems. It could be linear structures, where such an approach of connection as the common bus is used (Fig.4.2). On Fig. 4.2 four different units are shown (U1 – U4), which are connected by the common bus.

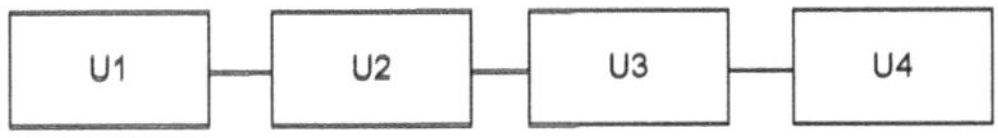

Fig. 4.2 Linear system structure with four elements

The system can have the radial structure (the star). It is typical for systems of wireless communications, where elements are connected with the same access point (AP), either a booster converter or other common node (Fig. 4.2).

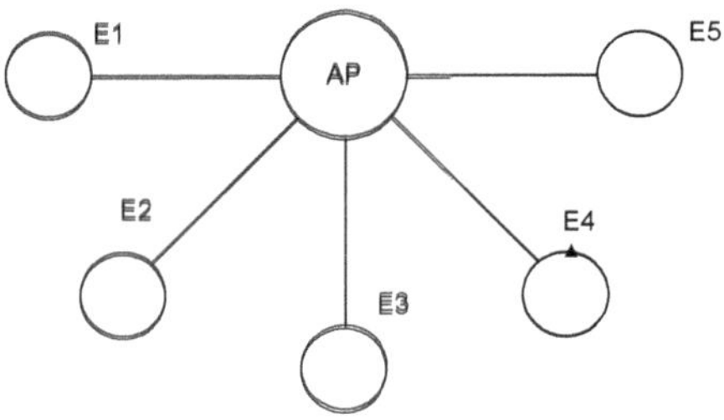

Fig. 4.3 Radial structure with common access point

In the case of n-level hierarchical system the radial-node structure is applied. In such a structure there is one common center controlling more than one radial or radial-node structure. Often, this organization is named tree-shaped. The networks of some institution, corporation, or bank can be mentioned as the examples of tree structures. The three-level radial-node network is shown in Fig. 4.4.

Of course, all these structures can be mixed together. It leads to the arbitrary (mixed) structure, and it is very difficult to reduce it to some known typical structure.

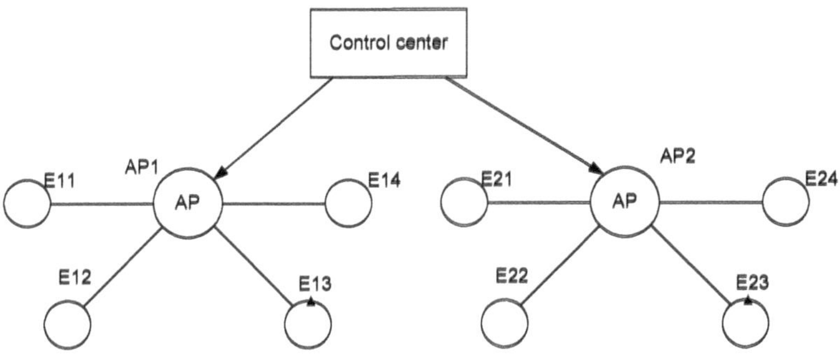

Fig. 4.4 Three-level structure of controlled network

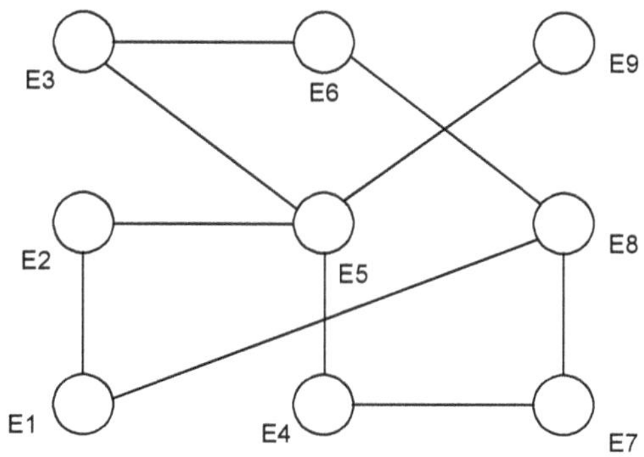

Fig. 4.5 Mixed structure of controlled network

There are two main types of controlled systems, namely single-element and multi-element systems. Let us discuss their characteristics in details.

The system is named a single-element system if it includes only one controlled element. Such a system can be either one-dimensional or multidimensional. In the case of one-dimensional system, it has only one state, $x(t)$. Multidimensional systems have many states, $\mathbf{x}(t) = (x_1(t), x_2(t), \ldots, x_n(t))^T$, one of these states, $x_i(t)$ can be controlled.

There are two principles used for constructing control systems, namely the principles of Ponselle and Watt. These principle are general, they do not depend on the type of a controlled system (situational, declarative, optimal criteria).

The Ponselle's principle is a control by influence. This principle is based on assumption that any discovered influence (failure, external influence, and need in some resource) always finds an adequate control as a response for this influence. It means that some backup device is introduced instead of defective, or the external influence is deleted, or some additional resource is used. The structure of the control system based on the Ponselle's principle is shown in Fig. 4.6. If applied control procedure does not possess high reliability, then some feedback information is organized from the site of controlled object. This feedback information confirms fulfilment of a control order.

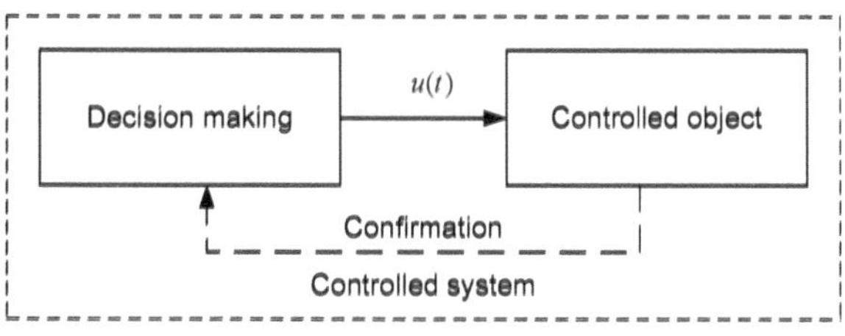

Fig. 4.6 Control system based on Ponselle's principle

This control principle is widely used in TCS. For example, all control information in the packet headers implements this very principle. When some PMD gives its order, it belongs to the class of control by Ponselle's principle, too.

The Watt's principle is a control by deviation (or control with the feedback). This principle is used in devices where input signals can have some deviations from their average (or standard) values. These controls are used in watching systems (trackers), in devices for stabilization of some output level, in automatic devices controlling the modes of differing network elements. Per se, the Watt's principle is the base for constructing automatic control systems (ACS). Let us remind that the principle of feedback is one of the most important principles used for organization of control. The structure diagram of control unit based on Watt's principle is shown in Fig. 4.7.

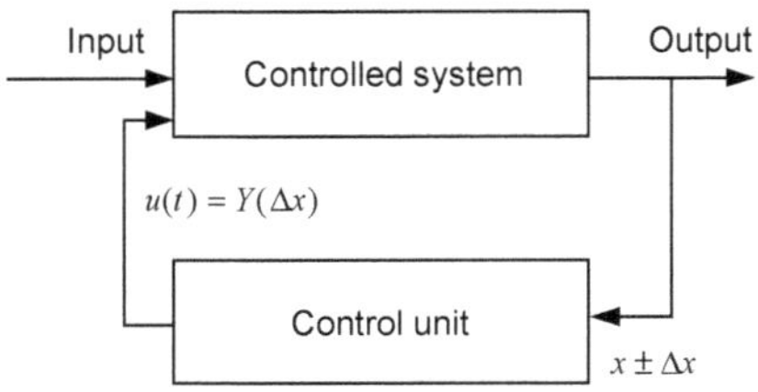

Fig. 4.7 Organization of control by Watt's principle

In this case, an operator (PMD) can be included into the control loop. For example, the operator can observe the evolution of some process ($x \pm \Delta x$) and control its progress after analysis of the value of deviation Δx. But the participation of a human being leads to considerable growth of lag effect (inertance) for the control loop. In contrast, the automatic control provides implementing requirements for performance, accuracy, and other important characteristics of a system.

Each from these principles of control differs in accuracy of task's accomplishment. The Ponselle's system executes control without deviations if all procedures are performed exactly. This property of controlled system is named statism.

In the same time, the Watt's principle is characterized by existence of deviation in the control loop, $(x+\Delta x)$. The value of Δx is named post-tuning drift (or residual detune). It is necessary for keeping the required state of a system applying some control $u(t) = Y(\Delta x)$. Obviously, removing the control $u(t)$ results in the loss of controllability for the controlled system $S(x,u,t)$. Therefore, permanent existence of the residual detune $\pm\Delta x$ is a factor, which should be taken into account in the systems with feedback. The systems with existence of the residual detune are named astatic.

Despite the used control principle, control systems can be either deterministic or stochastic (situational or optimal). It means that principles of Ponselle and Watt are general and can be applied in any control system.

As a rule, multi-element controlled systems are multidimensional. But there is one specific class of controlled systems having one common control element, whereas the number of control centres can be arbitrary (more than one). These structures are modelled by methods of the game theory. They will be discussed a bit later.

The multi-element controlled systems can be constructed using different methods. The most characteristic structures of these systems are the hierarchic centralized controlled systems (Fig. 4.8) and decentralized controlled systems (Fig. 4.9).

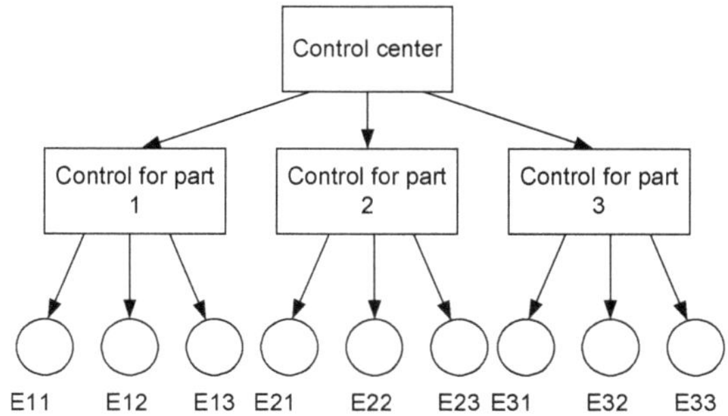

Fig. 4.8 Structure of three-dimensional hierarchic system

The centralized control methods can have two or more levels (striations, echelons). The top level (the control center of the system) controls some from underlying objects or centres. Each of these elements controls some underlying parts of the system and so on. Therefore, each upper level delegates authorities to underlying levels. Due to it, the underlying levels make the control center free from some routine operations. It increases the efficiency and quality of control.

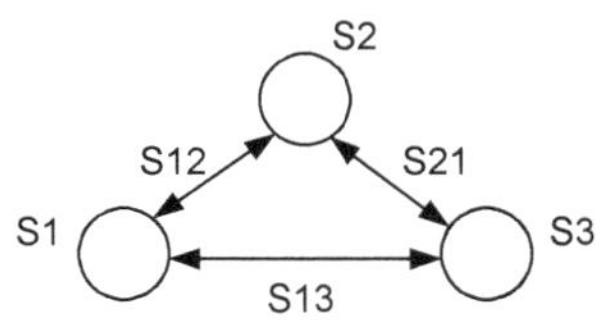

Fig. 4.9 Structure of three-dimensional hierarchic system

The hierarchic centralized control systems have the following features:

1. There is the possibility of distribution control functions and the tasks of control decision making among different control levels. The higher levels make solutions of strategic problems, whereas the solutions for tactic tasks belong to the underlying levels. It provides operability of decision making, as well as their higher accuracy.
2. The control elements for each level are autonomous. It means that each level is responsible for control decisions in the range of its authority.
3. There is the danger that some subsystem can act to the prejudice of the general goal of the total system. It is possible when such a subsystem is looking for achievement of its own goal.
4. The subsystem of higher level can have incomplete information about goals and constraints of underlying subsystems. It can decrease the quality of control.

The reasonable division of the general system on subsystems is an important factor under constructing hierarchic systems. It is necessary avoiding the control over instance. In this case, some elements are in situation of double submission (subordination). It results in failure of control, appearance of conflicts and cul-de-sacs in the system functioning.

There are two main advantages of hierarchic controlled systems: simplicity of control functioning; stability of controlled systems for applying both situational and formalized methods.

There are some drawbacks which are characteristic for the hierarchic controlled systems. The following of them are the most important: inertance, delays, lack of adaptability. The inertance is connected with the lack of efficiency under the solution of problems due to necessity passing through all control levels (bottom up and vice-versa). The delays in different elements of a network result in delay of signal and control information for controlled devices. It results in general delay of control in the system. These systems cannot be adapted automatically to the changes (variations) of input influences. Existence of these drawbacks results in inefficient usage of network resources.

If each element of a system can chose its own control decision, then such a control belongs to the decentralized control methods. These methods can be applied if a decision of any element is coordinated with other interrelated elements. The model of decentralized system is shown in Fig. 4.9. Here the interrelations are denoted as S_{ij}. If interrelations S_{ij} are very weak (or zeroed), the element S_{ij} uses its own recourse for decision making. If the level of interdependency increases, then recourses other elements are used in cooperation with the element S_{ij} for decision making. Therefore, the principle of cooperative decision making is used in the decentralized control system.

The state of n-element controlled system can be defined as

$$d\overrightarrow{x}(t)/dt = A(t)\overrightarrow{x}(t) + B(t)\overrightarrow{u}(t). \tag{4.4}$$

In (4.4) symbols A and B stand for matrices having size $n \times n$. These matrices determine such parameters as: inertance of element i (a_{ii}); relative inertance taking into account delays in elements of connection and processing (a_{ij}); the control recourse given away by the element i in its interest (b_{ii}); the control recourse given away by the element j in the interest of element $i(b_{ij}$), and, at last, there is

$$u_i(t) = b_{ii}u_{ii} + \sum_{j=1}^{n} b_{ij}u_{ij}. \tag{4.5}$$

As a rule, some collections of control strategies are set up under the design of controlled systems. These strategies are used as responses for some determined situations, where the control provides fulfilment the objective function of given system. It is possible to have more than one strategy, therefore there is

$$\overrightarrow{\gamma} = \{\gamma_1, \gamma_2, \ldots, \gamma_m\}. \tag{4.6}$$

In (4.6) the symbol $\gamma_l = \gamma_l(\overrightarrow{x}(t))$ defines the behaviour strategy if there is an influence l. The strategy determines such a state $\overrightarrow{x}(t)$ of the system, where the maximum quality (gain) can be reached. This gain, $J_l(\overrightarrow{x}, \overrightarrow{u}, t)$is determined as:

$$J_l(\overrightarrow{x}, \overrightarrow{u}, t) = \int_0^T \gamma_l(\overrightarrow{x}(t))dt + \varphi_l(\overrightarrow{x}(T)). \quad (4.7)$$

In (4.7) the first member is a current gain, and the second one is the final gain.

Therefore, the gain for each member of coalition is determined by the state for each member, as well as by the states of all members.

The main advantages of decentralized control methods are the following. Firstly, there is a possibility for placement of control units in the minimum distance from the controlled objects. It minimizes the general delay in the control loop. Secondly, the operability of control is increased, as well as the control quality. Thirdly, it is much easier to take into account the influences from environment, as well as influences of adjacent elements of the system. At last, it is simpler to adjust (adapt) the control to variability of situation.

Of course, the decentralized control methods have some disadvantages, too. Firstly, the control system is complex, because it is necessary to take into account influences of both the environment and all interrelated elements of the system. Secondly, there are significant losses of network throughput for transferring service information among some control units and controlled objects. Thirdly, there is danger for arising unstable modes in the control system, as well as conflicts, deadlocks, closed cycles, and so on.

The mixed control structures are often used in practice, where centralized systems have some elements of decentralization.

Multi-element control methods assume existence of more than one control centres, where each of them is connected with other. These centres can have goals, which are coincident, partially coincident, or opposite (antagonistic). The particular interest represents multi-element control methods having single mutual controlled element. The similar problems can be solved using the methods of game theory. In this theory there are tasks with opposite interests and tasks with cooperative interests. The following equation represents the state of the system with many control elements:

$$dx(t)/dt = ax(t) + \sum_{i=1}^{n} b_i u_i. \quad (4.8)$$

In (4.8) the symbol u_i represents the control influence number i from the control center i, the symbol b_i stands for element of the vector $b = b_1, b_2, \ldots, b_n$, this element depends on both the state $x(t)$ and other influences. Therefore, the control in the system is executed taking into account the vectorial criterion

$$J_i(x, \overrightarrow{u}, t) \rightarrow \underset{u,t}{extr}. \quad (4.9)$$

It is possible to have some variants of the task (4.9) in dependence of interests of each control center. Firstly, the cooperative solution can be found. In this case the general gain is maximized for all control centres:

$$J(\overrightarrow{x}, \overrightarrow{u}, t) \rightarrow \max_{u,t}. \tag{4.10}$$

It means that all centres work to reach the general result and no center has interest in diminished gain for other centres. In this case, the strategies of each gamer can be either inter-dependable or undependable. The conflict solution (the antagonistic game) appears when there are conflicts among the interests of gamers. It can be represented by the following equation:

$$J_1(x, u_1, t) = -J_2(x, u_2, t). \tag{4.11}$$

This equation shows that the gain of one gamer means the loss for other participant of the game. Such a system having two antagonistic gamers is shown in Fig. 4.10.

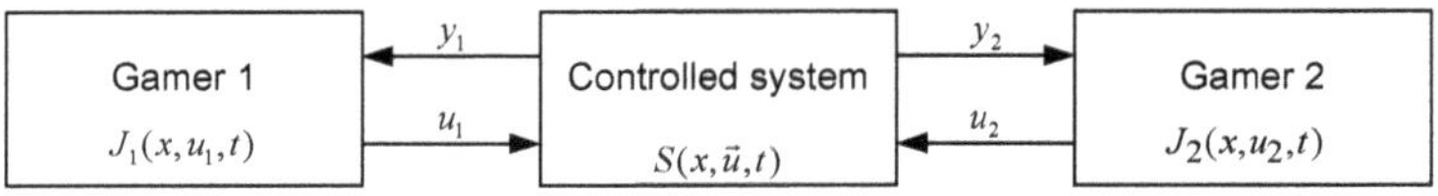

Fig. 4.10 Controlled system having two antagonistic gamers

If there are n gamers, the antagonistic game is reduced to formation of coalitions. In the common case, if there are two antagonistic gamers (x_1 and x_2), then it is necessary to make corresponding efforts $u(t)$, interpreted either as the control or as the payoff. In this case the gamer 1 maximizes his payoff, whereas the gamer 2 minimizes its gain. The strategies of behaviour for these gamers can be written as:

$$\gamma_1 \in P_1(u) \rightarrow \inf_{x_2} u(\gamma_1, x_2) = \sup_{x_1} \inf_{x_2} u(x_1, x_2);$$

$$\gamma_2 \in P_2(u) \rightarrow \sup_{x_1} u(x_1, \gamma_1) = \inf_{x_1} \sup_{x_2} u(x_1, x_2),$$

where $\sup y$ and $\inf y$ means correspondingly the upper and down boundaries of the function y. The numbers $\sup \inf u$ and $\inf \sup u$ represent correspondingly guaranteed gain of the gamer 1 and guaranteed loss for the gambler 2. They are connected by the following inequality:

$$\sup_{x_1} \inf_{x_2} u \leq \inf_{x_2} \sup_{x_1} u. \tag{4.12}$$

The inequality (4.12) can be explained by the fact that the maximum gain of the first gamer cannot be exceed the lost of the second gamer. The quantity α (the price of the game) can be found from the equation (4.12):

$$\sup_{x_1} \inf_{x_2} u = \inf_{x_2} \sup_{x_1} u = \alpha.$$

Conflicted decisions can be obtained for both deterministic (decision for the pure strategy) and stochastic (decision for mixed strategy) models. It can be shown that if there are no solutions for the pure strategy, then it is always possible to find a solution for mixed strategy.

There are static and dynamic (differential) games. In the case of the static game only single numerical solution can be found. In the case of differential game the equation (4.11) is transformed to the following one:

$$\Phi(x, dx(t)/dt, u_1, t) = -\Phi(x, dx(t)/dt, u_2, t), \tag{4.13}$$

As a result of solution for a differential game, the optimal strategies for both gamers are found. It leads to the point of equilibrium (the Nash's point). There is no sense for any of gamers to depart from the Nash's point.

There is a strategy of indifference or the game against nature. In this game no gamer has interest to its outcome. Thus, one of the criteria in strategies (4.11) and (4.12) is constant or equal to zero. Obviously, this task is reduced to the simple problem of stochastic control with interferences having the nature of the white Gaussian noise. Therefore, the usual control problems are only particular cases of game problems.

4.3 Mathematical Models of Deterministic Control

The control $u(t)$ is a dynamic procedure developed in time. Using this procedure, a controlled system, $S(x,u,t)$ is transferred from the state S_1 into the state S_2. This transition obeys to some rule (criterion), that is, the transition can be executed either during the minimum possible time or with minimum deviations from the optimal trajectory, or with minimum discharge of resources and so on. Besides, the model of a system should be created as a state equation.

The state equation of a controlled system can be written as the following one:

$$dx(t)/dt = Ax(t) + Bu(t). \tag{4.14}$$

In (4.14) the symbol A stands for the coefficient determining inertance of the system (or its performance), the coefficient B determines the degree of influence for the control signals $u(t)$.

If there is no control $u(t)$, then the system (4.14) turns out in the free, autonomous described by the following state equation:

$$dx(t)/dt = Ax(t) + B. \tag{4.15}$$

If system (4.15) is removed from its equilibrium state to the level B, then it will go back to this state in correspondence with the value A. The trajectory of the system is determined by the following equation:

$$x(t) = Be^{-At}. \tag{4.16}$$

This equation determines some possible trajectories for coming back into the equilibrium state. They are shown in Fig. 4.11.

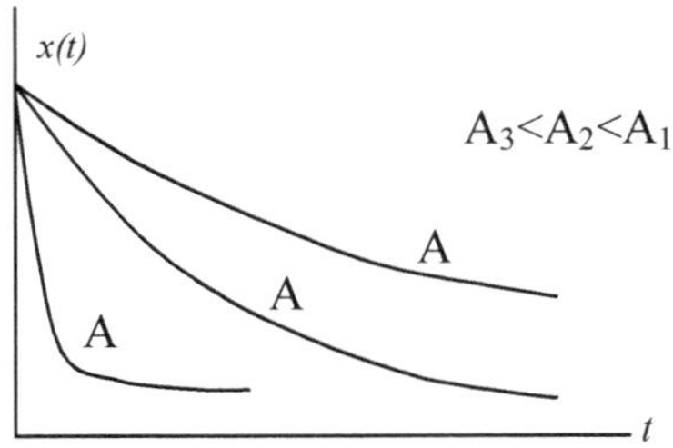

Fig. 4.11 Trajectories of system with different A

To operate in the stable mode, it is necessary to have $A < 0$. If there is $A > 0$, then the chaotic (catastrophic) mode takes place and the system is in the state of auto-excitation.

The mathematical model of a multidimensional controlled system is represented as a system of differential equations. An example of such a system is represented by the system (4.16).

$$\begin{cases} \dot{x}_1 = a_{11}x_1 + a_{12}x_2 + \ldots + b_{11}u_1(t) + b_{12}u_2 + \ldots + b_{1n}u_n, \\ \dot{x}_2 = a_{12}x_1 + a_{22}x_2 + \ldots + b_2u_2(t) + b_{22}u_2 + \ldots + b_{2n}u_n, \\ \cdots\cdots\cdots\cdots\cdots\cdots\cdots\cdots\cdots\cdots\cdots\cdots \\ \dot{x}_n = a_{1n}x_1 + a_{n2}x_2 + \ldots + b_nu_n(t) + b_{n2}u_n + \ldots + b_{nn}u_n. \end{cases} \tag{4.17}$$

Equations (4.17) represent n-dimensional system having mutually connected components. These connections are represented by state coefficients, a_{ij} and controls, b_{ij}.

In this system the coefficients a_{ij} and b_{ij} represent connections between elements i and j. The system will disintegrate by n undependable one-dimensional systems, if there are $a_{ij} = 0$ and $b_{ij} = 0$ (where $i \neq j$). The degree of these connections can be changed by control influences in the tasks of routing, commutation, and reconfiguration.

In the system theory, the systems are named degraded if they lost interconnections (it is a system with progressive factorization). If degrees of interconnections increase, then it corresponds to systems with progressive systematization. For these systems, there is a growth for their integrity and emergence. It is accompanied by appearance of new integrated properties, the quality and quantity of these properties is directly proportional to the existence of interconnections. Therefore, it can be affirmed that exactly the interrelations a_{ij} and b_{ij} between the elements of separate subsystems $x_i(t)$ give some super-integrated properties, which are not the result of simple summation for the properties of separate elements or subsystems. It gives the property of integrity to the whole system.

It is convenient to show the states of a system by some trajectories on the phase plane (Fig. 4.12).

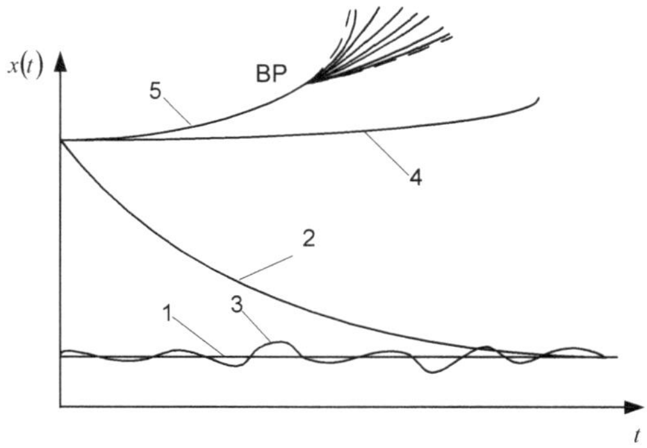

Fig. 4.12 Trajectories of dynamic systems with different states

There are five different states shown in Fig. 4.12, namely: equilibrium (1), transitional (2), stationary (3), critical (4), and catastrophic (5). Let us analyse these states.

The equilibrium state is an autonomous state or a state of rest when there are no external influences. The following equation represents the state of rest for some system

$$dx(t)/dt = 0. \tag{4.18}$$

If a system is in the transitional state, it will come back into the equilibrium state (after the external influence). The following equation represents the state of such a system

$$dx(t)/dt = Ax(t) + B. \tag{4.19}$$

A system is in the stationary state when the external influences are stationary too. The following equation represents the state of rest for some system

$$dx(t)/dt = Ax(t) + Bu(t). \tag{4.20}$$

In (4.20) the symbol $u(t)$ stands for an external influence. There is no stationary state for unstable systems. They transit into critical or catastrophic states.

The critical state is a boundary between stable and unstable states of a system. If there are further external influences, then the state trajectories will acquire the multiple natures. The state equation for this case can be similar to (4.20), where there is $A \approx 1$. If there is $A > 1$ in the equation (4.20), then a system is in the catastrophic state. It is a state of chaos. The next state of the system after the bifurcation point (BP) can be arbitrary one. But it is in the area restricted by two dotted envelopes.

In practice, a system state depends on such parameters as the coefficient A, the coefficient B, the value of the step of digitalization Δt, and some other parameters. Let us analyse these dependences.

4.4 Choice of Control Criteria and Finding Solution for Optimal Control

A criterion determines the quality of control. There are no absolutely (perfect) optimal controlled systems because there are no absolute criteria. There are optimal systems for one criterion or for some group of quality indexes. The one-parameter criteria are used very often (for example, minimum or maximum either of time, or price, and so on). It is very difficult to deal with multi-parameter criteria (for example, minimum cost and maximum performance). As a rule, the corresponding decisions are rather subjective, because the importance of parameters is determined by PMD. Moreover, many parameters are in the state of antagonism. For example, there is antagonism between the quality and price, or between performance and accuracy.

The most important requirements for criteria are maximum simplicity and direct connection with the essence of a control problem. The most spread are the following three classes of criteria: the criterion of suitability, the optimality criterion, and the criterion of global advantage.

The criterion of suitability is a rule allowing treating a system $S(t)$as a suitable system if all quality indexes J_i correspond to some acceptable or specified values. It corresponds to the following equation:

$$J_i \geq J_{\mathrm{acc(i)}}, \forall i. \tag{4.21}$$

The criterion of suitability is used widely in TCS. The known criterion of service quality, QoS belongs to these criteria. This criterion is named the principle of guaranteed quality, where the value of $J_{\mathrm{acc(i)}}$ is determined by requirements of QoS.

The optimality criterion is a rule allowing treating a system as optimal according with the chosen quality index

$$J_i = \max, \forall i. \tag{4.22}$$

Using this criterion, the automatic control algorithms operate. As a rule, either some quality is maximized or some loss is minimized.

The criterion of global advantage is a rule, when a system is optimal for all possible quality indexes. Such a situation of global optimization arises very rarely, because, as a rule, different criteria are antagonistic to each other.

Obviously, the efficiency in reach of desired quality is different for different criteria. The highest quality is reachable for the criterion of global advantage, whereas the lowest for the criterion of suitability.

One of the most used criterions for optimal control is a criterion of minimum mean-square deviation (MMSD). This criterion minimizes average losses for power, cost, and quality and so on. It can be expressed by the following equation

$$J(x,u,t) = \frac{1}{2}x^2(t_F)D + \frac{1}{2}\int_0^{t_F} \left[x^2(t)Q + u^2(t)R\right] dt \to \min. \tag{4.23}$$

In (4.23) the symbols D and Q stand for coefficients determining minimum loss for some finite controlled part and along the total trajectory correspondingly. In turn, the coefficient R determines minimum loss for control actions (minimum loss of energy, fuel and so on).

Let us discuss how to find a solution of optimization task for finding control in case of deterministic systems. Let us substitute the value of given system's state $x(t)$into the equation for criterion $J(x,t)$. As a result, the following value of $u(t)$can be found

$$u(t) = -R^{-1}BP(t)x(t). \tag{4.24}$$

In (4.24) the value $P(t)$ determines the mean-square deviation for the system's state during the movement along the optimal trajectory. The value of parameter $P(t)$ is determined after finding solution of the differential equation of Riccati:

$$dP(t)/dt = -2P(t)A + P^2(t)B/R - Q. \tag{4.25}$$

To find the control error in the stable state, it is necessary to make the equality $dP(t)/dt = 0$ and obtain values of $P(t)$ for $t \to \infty$.

The block diagram of optimal control algorithm is shown in Fig. 4.14. This algorithm corresponds to equation (4.24). As a rule the exact values of parameters of a controlled system are not known in many practical cases. It leads to divergence of these parameters with parameters of a chosen model. It results in deviations in the trajectory of optimized system. These deviations are interpreted as different modes.

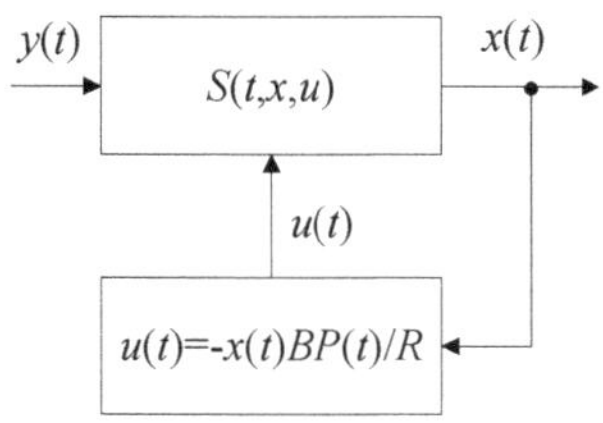

Fig. 4.13 Block diagram of optimal control algorithm

The control modes used in the transition period are determined by inertance of a system and by degree of control influence. Three possible trajectories for movement of a system are shown in Fig. 4.14. The system transfers from the state $x(t_0)$ into the state $x(t_F)$.

In the case of optimum mode a controlled system reaches its equilibrium state moving along the optimum trajectory in minimum time, $t_{\min}$. The mode of undershoot is a suboptimum mode when there is $B < B_{opt}$. The equilibrium state is reached in the time $t > t_{\min}$. The mode of overshoot is a suboptimum mode when there is $B > B_{opt}$. The system reaches its equilibrium state operating in the oscillation regime. In this case there is either $B \gg B_{opt}$ or $A \approx 1$.

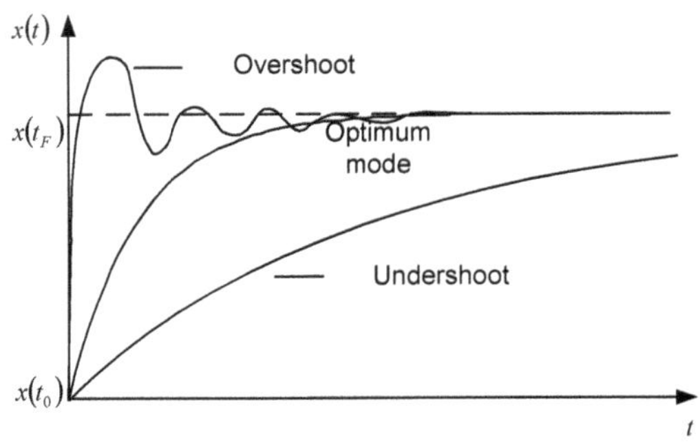

Fig. 4.14 Trajectory of controlled system

Other approaches are used in the case of stochastic systems. In this case a state $x(t)$ is simulated by a random process. The state equation of a stochastic system is represented as

$$dx(t)/dt = Ax(t) + Bu(t) + C\xi(t). \tag{4.26}$$

In (4.26) the function $\xi(t)$ determines a virtual generated process having the type of the Gaussian white noise, the coefficient C determines the level of this noise and, therefore, of the process $x(t)$. Other coefficients and variables in (4.26) have the same meaning as in equation (4.4) for deterministic systems.

Trajectories of random states $x(t)$ differ from the ones shown in Fig. 4.14. The deterministic transitional process (Fig.4.14) is mixed with some random process $v(t)$, which is present in the observer equation (4.1). This random process $v(t)$ permanently leads the system out from its equilibrium state (Fig. 4.15).

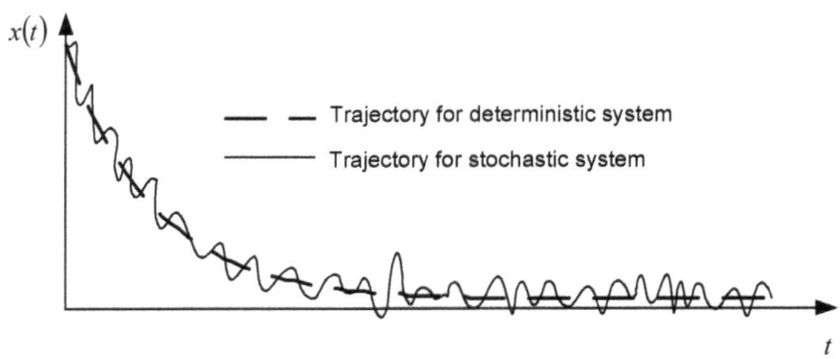

Fig. 4.15 Differences in trajectories to equilibrium state

Any deterministic criterion cannot be used for finding control of stochastic system. It is connected with the fact that current random values for $x(t)$ and $u(t)$ from equations (4.22) and (4.23) are unknown and random. In this case, the average value of a criterion is used. It results in the following equation for the chosen criterion:

$$M\left[J(x,u,t)\right] \to \min. \tag{4.27}$$

For finding optimal value of (4.27), it is necessary to substitute the random process $x(t)$ into criterion; it makes the finding solution more difficult. Because of it, they do not look for immediate value $u(t) = \varphi(x(t))$, but use the results of decomposition (or separation) theorem.

This theorem is formulated as the following one. Let it be the Gaussian state $x(t)$ and let the MMSD criterion be used. In this case the optimal control for a random system $S(x,u,t)$ can be executed as two separate procedures. One of them is the optimal stochastic estimation of the state $\hat{x}(t)$ and the second is the procedure of deterministic control $u(t) = \varphi(\hat{x}(t))$.

The block diagram of controlled system based on the results of the decomposition theorem is shown in Fig. 4.16. The deterministic procedure for control is found from (4.24). But now the estimation $\hat{x}(t)$ is used instead of $x(t)$. The procedure of optimal stochastic estimation is found as a function from observation (4.1): $\hat{x}(t) = f(y(t))$. These algorithms includes procedures of stochastic approximation by Robbins – Monro, Kiefer – Wolfowitcz, Newton – Raphson, filters Kalman – Busy, methods of Markov's nonlinear filtration and so on. The block of deterministic control is represented by the equation $u(t) = Y(\hat{x}(t))$.

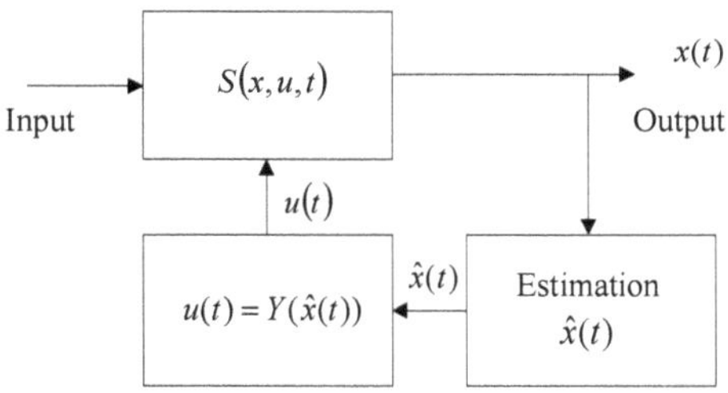

Fig. 4.16 Block diagram of controlled system based on the results of the decomposition theorem

Therefore, the main problem in finding the optimal control $u(t)$ consists in finding estimates $\hat{x}(t)$. Besides, the controlled systems should satisfy to some requirements, such as stability, observability, identifiability, adaptability and so on.

Recommended Literature

1. Dorf, R., Bishop, R.: Modern control systems, 11th edn. Pearson Prentice Hall, London (2007)
2. Kwakernaak, H., Sivan, R.: Linear optimal control systems. John Wiley and Sons, Chichester (1972)
3. Tanenbaum, A.: Computer Networks. Prentice Hall PTR, Englewood Cliffs (2002)

Chapter 5
Methods for Providing Controllability of TCS

Abstract. The chapter is devoted to the methods of providing controllability. Some conditions are discussed, such as observability, identifiability, stability, and invariance. Attention is paid to the methods of sample statistics under the assumption of observation of random variables, random processes and random fields, as well as sample parameters for different values of correlation windows. The recommendations are given for construction of sample and recursive estimates, used for different random objects, as well as for different criteria of their optimality.

5.1 Methods for Providing Observability

The controllability is a property of a system to transit from one state into another in a required time and along a required trajectory. This property is achieved if any component of the state $x_i(t)$ of the system $S(t,x,u)$ is accessible for control actions (influences) $u(t)$. To provide the controllability, the conditions of "reachability" should be satisfied. This property can be interpreted as the ability in transition into these or those states under existence of restricted resources. The property of controllability is present if the following conditions are present: observability, identifiability, stability, and invariance. Let us analyse these conditions starting from the property of observability.

The observability is provided if there is a possibility of obtaining information about each component of the state vector $\overrightarrow{x}^T(t) = x_1, x_2, \ldots, x_n$. If the observation (measurement) can be executed without errors, then the following deterministic equation can be obtained:

$$y(t) = Hx(t). \tag{5.1}$$

In (5.1) the symbol H is the gain coefficient (or reduction factor) under the measurement of $x(t)$.

V. Popovskij et al.: Control and Adaptation in Telecommunication Systems, LNEE 94, pp. 55–65, 2011.
springerlink.com

If there are either some errors in measurements or noises in the observation channel $v(t)$, then the equation (5.1) becomes the stochastic equation of observation:

$$y(t) = Hx(t) + v(t). \tag{5.2}$$

The values of $y(t)$ from equation (5.1) can be directly used for control. If the equation of observation is stochastic, then $y(t)$ is a random process. To use it in a control algorithm, it is necessary to estimate its state $\hat{x}(t)$. In the real conditions of observation, the equation (5.2) is a sample from the results of assessment:

$$\overrightarrow{y}^T = y_1, y_2, \ldots, y_n. \tag{5.3}$$

Therefore, the equation of observation becomes discrete and it is represented as the following:

$$y_k = Hx_k + v_k. \tag{5.4}$$

The discrete sequence of observation should be processed using, for example, the estimation procedure. Next, these results can be used in a control algorithm.

5.2 Methods of Identifiability

The identifiability is ability (possibility) of constructing some mathematical model of a system and obtaining information about properties of this simulated system on the base of results of observation. In other words, it is possibility to overcome a-priori uncertainty about these or those system properties of a system model using the results of processing of observation. For example, the quantities of dispersion, mean value, correlation interval are unknown in many practical cases. In this case the problem of identifiability arises for one or more parameters. It means that these parameters should be measured. There are two methods of estimation. The first of them is a sample estimation. It can be obtained from the formula used for calculating mean value, $x_{cp} = \sum_{i=1}^{n} x_i p(x_i) dx$. If probabilities of all events are equal ($p(x_j) = p(x_i) = 1/n$), then we can get the sample mean:

$$\hat{x}_{cp} = \frac{1}{n} \sum_{\text{k}=1}^{n} x_k, \tag{5.5}$$

where n is sample size.

Acting in the same manner, it is possible to get the calculation formula for sample estimate of dispersion:

$$\hat{\sigma}^2 = \frac{1}{n-1} \sum_{\text{k}=1}^{n} (x_k - \hat{x})^2. \tag{5.6}$$

In (5.6) the part $\hat{\sigma} = \sqrt{\hat{\sigma}^2}$ determines the mean-square deviation.

Therefore, the sample estimates (5.5)-(5.6) are obtained after some time period, which is necessary for collecting the full statistics and execution of calculations. In some cases, tremendous delay in obtaining the sample estimates does not permit its use for control problems.

The recursive estimation assumes processing the sample statistics in the real time, step by step. Each step of recursion is connected with obtaining a current sample value y_k. This value is compared with the estimate from the previous step $(y_k - \hat{x}_{k-1})$. Next, it is added to already obtained estimate with some weight $1/\gamma$. As a result, the following value can be found:

$$\hat{x}_k = \hat{x}_{k-1} + \frac{1}{\gamma}(y_k - \hat{x}_{k-1}). \tag{5.7}$$

According to the decomposition theorem both estimates (5.5) and (5.7) can be used for control. But usage of sample estimates (5.5) under automatic control methods leads to delay in response for these or those influences. In the same time, these estimates are used in the situational control methods, where the response of an operator leads to delays in decision making. In contrast with sample estimates, the recursive estimates (5.7) are calculating for each control step and it allows making decisions in the real time. Obviously, there are some constraints for use of both approaches. Let us discuss them in details.

There is the following peculiarity in the sample statistics. As a rule, it is formed as some discrete simulations (discrete observations)

$$y(k) = Hx(k) + v(k), k = 1, 2, \ldots, n. \tag{5.8}$$

If there is $H = 1$, then we have the following equation

$$y(k) = x(k) + v(k). \tag{5.9}$$

It means that there is a typical situation of observation of some random process (or random value) $x(k)$ on the phone of interferences $v(k)$ having a nature of Gaussian white noise. The random nature of observations $\overrightarrow{y}^T = y_1, y_2, \ldots, y_n$ is stipulated by both the randomness of the observed object $x(t)$ and actions of noises $v(t)$ in the observation channel. The randomness of the observed object $x(t)$ can have different sense. It means that the sample statistics should be formed in different ways.

To get the sample statistics, it is necessary to make the following actions. Firstly, it is necessary to determine the class to which an observed object $x(t)$ belongs. At least, two variants are possible, namely the object $x(t)$ is either a random value or a random process. The further steps for forming the sample (5.8) depend on this choice.

If it is assumed that the object $x(t)$ is a random process, then it has constant values $(x(t) \rightarrow const)$ during the interval of observation $T = t_1, t_2, \ldots, t_n$. But the value of $x(t)$ (or $x(k)$) is unknown. Moreover, it is random. As a result, the reading values y_k have some dispersion and there is $y_k \pm \Delta y_k$. This dispersion is determined by both the noise and measurement errors. For any organization of a sampling process, the reading values of y_k are independent because of noncorrelatedness of the values of noise v_k. The obtained sample sequence $y_k = x_k + v_k$, $k = 1,2,\ldots,n$ should be processed using (5.5) to get the values of mean $\hat{x}$ or other characteristics of a random value x. If there is a reason to treat $x(t)$ as a random process, then $y(t)$ is changed due to alterations of the process $x(t)$ and because of the noise $v(t)$.

To find the sample statistics for the process $x(t)$, it is necessary to choose two main parameters (Fig.5.1): the observation interval T_o and the sampling rate $f_d = 1/\tau_d$, where the value τ_d is an interval between two adjacent samples t and t_{-1}. The choice of these parameters depends on the further use of the statistics $y_1, y_2, \ldots, y_n$.

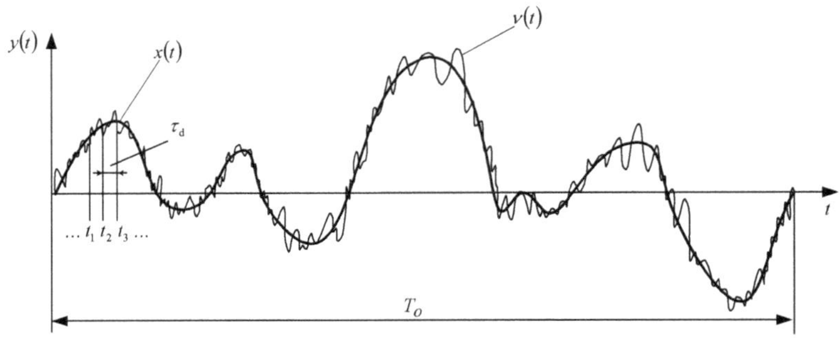

Fig. 5.1 Implementation of observed random process $y(t)$ for interval T_o

If the statistics is used for obtaining sample estimates (5.5) for the mean value $\hat{x}_{cp} = n^{-1}\sum_{K=1}^{n} x_K$, then the interval between reading values should be chosen from the conditions of independence between the adjacent readings. It means that the correlation coefficient should go to zero ($r_{ij} \to 0$):

$$r_{ij} = \frac{\sum_i^n \sum_j^n (x_i - x_{cp})(x_j - x_{cp})}{\sqrt{\sum_{i=1}^n (x_i - x_{cp})^2 \sum_{j=1}^n (x_j - x_{cp})^2}} \to 0. \tag{5.10}$$

Let us discuss some methods used for forming such a sample. Let us start from methods oriented on obtaining independent sample values.

The Shannon–Kotelnikov' theorem is used in practice for obtaining independent readings. According to this theorem the intervals between the readings are chosen from the following conditions $\tau_d = (2F_{\max})^{-1}$. In this formula the value $F_{\max}$ is a maximum frequency of spectrum of the random process $x(t)$. Let us point out that the spectral distribution $S(\omega)$ and correlative function $R(t)$ are interconnected by the pair of Fourier transforms.

There is another approach for obtaining independent readings. It is based on the fact that the readings should be made through the interval which is more or equal to the correlation interval τ_{cor}. The correlation interval of the process $x(t)$ is determined as $\tau_{cor} = 1/\alpha$, where the symbol α stands for double-ended width of the spectrum for the estimated process. Obviously, there is the equality $\alpha = 2\Delta F$ for the double-ended spectrum. The correlation interval is equal to the phase shift of correlative function $R(t)$ for the interval where it is decreased on the value 0,37 from its maximum (Fig. 5.2). To do it, the correlative function is approximated by a double-side exponent:

$$R(t) = R_0 e^{-\alpha|\tau|}. \tag{5.11}$$

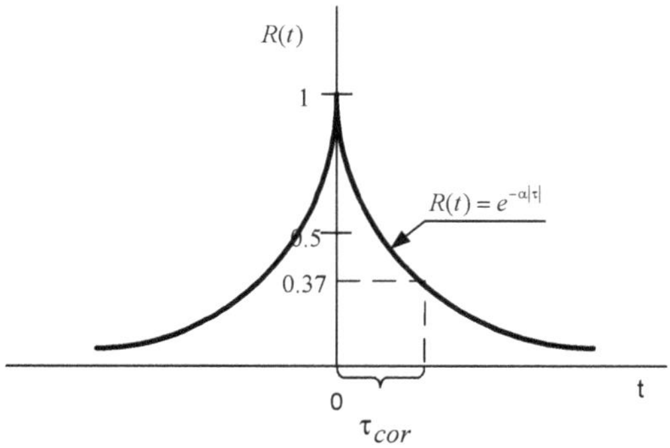

Fig. 5.2 Finding correlation interval

There is an effective frequency band F_{eff}. Numerically, it is equal to the base of triangle such that its altitude is equal to $S(\omega)$ max, whereas its area is the same as the area under the line $S(\omega)$. If the function $S(\omega)$ possesses the characteristic of the normal law with $\int_{-\infty}^{\infty} S(\omega)d\omega = 1$, then there is: $F_{\mathrm{eff}} = S_{\max}^{-1}$.

The diagram shown in Fig. 5.2 represents function (5.11) for the case $R_0 = 1$. The value $\tau_{cor} = 0.37$ is not absolute. It is calculated from formula (5.11) under the condition that there is $\alpha\tau = 1$. This equality is reached if there is $R(t) = e^{-1} = 0,37$. It is assumed in engineering practice that if correlation is less than $R(t) < 0.37$, then such objects (or their sections such as random values, processes or fields) should be treated as uncorrelated.

Comparison of these two approaches shows that the results of methods for choice the independence of readings are identical. It follows from the theorem of readings that there is $\tau_d = (2F_{\max})^{-1}$, whereas the value for alternative method is determined as $\tau_d = \tau_{cor} = 1/\alpha = 1/2\Delta F$. This conclusion is justified by the equality $F_{\max} = 2\Delta F \approx F_{\mathrm{eff}}$.

The following reasons explain the necessity for independent readings of samples. Let us presume that the readings are strongly dependent (Fig. 5.3). Let the process $x(t)$ change very slowly and let the observation interval (T_o) be commensurable with the correlation interval (τ_{cor}). It gives the sample estimate $\hat{x}_{cp}$, which is displaced. It means that there is $\hat{x}_{cp} \neq 0$, but in reality there is $x_{cp} \approx 0$.

But if the readings are made rarely (when there is $\tau_d \gg \tau_{cor}$), some important information about a process can be lost. For example, the real process showed in Fig. 5.4 changes quickly enough. In the same time, the readings are executed too rarely (in the instances t_1, t_2, t_3, t_4). It results in absence of correspondence between the real process and its approximative function shown by the dotted line. The specific outliers of the process $x(t)$ are not recognized by the approximative sample in use.

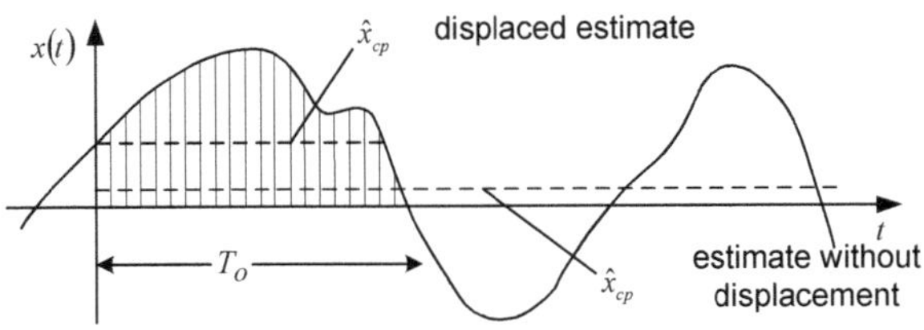

Fig. 5.3 Obtaining mixed estimate for correlative samples

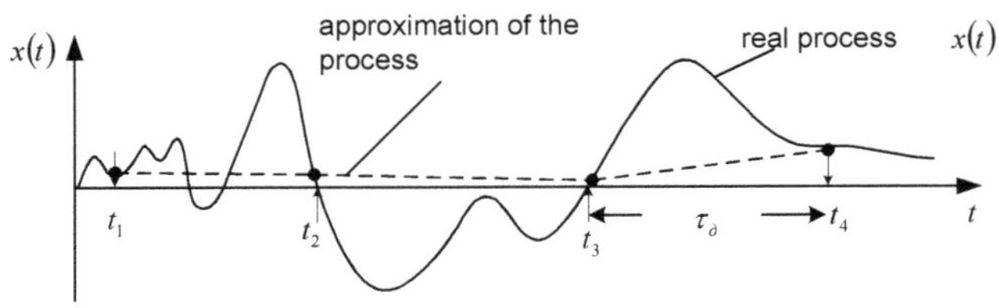

Fig. 5.4 Example of unreasonable rare sample

It is clear, that the estimate $\hat{x}_{cp}$ can be treated as efficient and unbiased $((\hat{x}_{cp} - x_{cp}) \to 0)$ only in the case of stationary ergodic process $x(t)$. If the process is not pure stationary, then the displacement (trend) Δx of the estimate appears:

$$\hat{x}_{cp} = \hat{x}_{cp} \pm \Delta x.$$

Therefore, the sample estimates cannot be used to characterize non-stationary processes.

The sample size n should be maximally large and the following inequality should take place: $T_o \gg \tau_{cor}$. It allows evading the displacement effect shown in Fig. 5.3. In the ideal case there is either $T_o \to \infty$ or $n \to \infty$.

These peculiarities should be taken into account when the monitoring of network elements is organized. It should be done for any management protocol in use (for example, for SNMP).

There are some specifics in formation of a sample for the case of recursive estimation. This kind of estimation is used when it is necessary to get the estimate in the real time. Recursive procedures perfectly match with random processes and corresponding control tasks. It is due to the fact that recursive procedures reflect the dynamic of their variations in time.

The recursive property is characteristic for Markov's processes. Thus, a state of Markov's process is represented by the following expression:

$$x(k+1) = \Phi(k+1,k)x(k) + G(k+1,k)\xi(k). \tag{5.12}$$

In (5.12) the part $\Phi(k+1,k) = e^{-\alpha\Delta t}$ determines the state coefficient corresponding to the coefficient A in continuous presentation. In this case there is $A = \alpha = 2\Delta F$, where Δt is the digitalization step (it is the time interval between the readings t_i and t_{i+1}). The coefficient $G(k+1,k) = \sqrt{\sigma_\xi^2(1-e^{-\alpha\Delta t})}$ takes into account the level of generating sequence $\xi(k)$, as well as the level of the process $x(k)$. The symbol σ_ξ^2 is a spectral density of the process power $\xi(k)$.

Analysis of the state equation (5.12) shows that the first summand ($\Phi(k+1,k)x(k)$) allows taking into account peculiarities of changing of predicted component of the process $x(k)$ for the next step. The function $e^{-\alpha\Delta t}$ is a function of prediction. It means that there is a large probability that the current value $x(k)$ will be changed by the value $e^{-\alpha\Delta t}$. Additionally, the more is the value α, the more these alterations are visible. So, it determines the level of performance for the process $x(k)$.

Therefore, the time interval between readings should be as short, as possible to reach the desirable accuracy of approximation. The following short conclusions can be made for the discussed methods of sample and recursive estimates. Firstly, some number $\hat{x}$ is obtained as a result of sample estimate of a random process $x(t)$ (or the quantity (5.5)). This number characterizes the mean value of estimated sequence for given observation interval. The sample readings $y_k, y_{k+1}, \ldots, y_n$ should be independent in the maximum possible degree. Their number should be big enough to provide the adequacy of sample and absence of bias in obtained estimates of the mean value of process $x(k)$.

Secondly, the sequence of estimated values $\ldots\hat{x}(k-1), \hat{x}(k), \hat{x}(k+1)\ldots$ is formed as a result of recursive estimate for a given function under observation. This sequence is named the conditional mean or a posteriori mean. The sample readings $\ldots y(k-1), y(k), y(k+1)\ldots$ should be inter-dependable in the maximum possible degree. It guarantees greater accuracy of consequential estimates.

Both kinds of estimates (sample and recursive) are widely used in the practice of communications. Let us discuss some examples for their applications.

1. They are used in control tasks when the decomposition theorem is used. The recursive algorithms of estimate are generally used under organization of automatic control. It follows from the fact that they provide obtaining the current value of observed state and this value is processed immediately for the control of a system. It provides minimum delay in the control loop. But if this delay is not critical, then the sample estimates can be used. For example, in the tasks of situational control, as a rule, these estimates are used. It follows from the fact that a PMD makes decision in this case.
2. These estimates can be used in the tasks different from the tasks of automatic and situational control. The examples of these tasks are the problems connected with formation of archives, displaying monitored information, under statistic tests, and for analysis of results of scientific researches.

5.3 Stability and Invariance of Control System

The term "stability" is used for determining the properties of both complex systems (such as TCS) and simple control systems. In the case of complex systems, the factor of stability is composite. It includes some components. The first of them is a reliability which determines the ability of execution of required function for a given time interval. Very often, this component is determined as the number of percents in execution of the required function. With regard to TCS, this component should be equal to $H \geq 99.9999\%$ of the required time. It corresponds to the loss percent equal to $T = 100 - H = 0.0001\%$.

The second component is named "survivability" and it determines the property for execution of the required function under given conditions. These conditions can be the elevated temperature, elevated humidity, existence of hostile environment (corrosive medium), and elevated seismicity and so on. The third component is named "external immunity" or "interference protection". It is invariance to this or that class of interferences.

The stability is comprehended in narrower sense when it is a stability of simple systems (or any dynamic system). Any dynamic system can transit into the unstable state (it could be even critical or catastrophic) if either its parameters or input actions were not chosen in a proper way.

Stable systems possess a restricted response under restricted input actions. It means that there is a constraint for the value of derivative of output for input:

$$dy(t)/dx(t) < \infty.$$

There are absolutely stable systems (total stability) when a system remains stable for any input actions. There are restrictedly stable systems, when they are stable in some restricted area of input actions.

Different factors of stability can be found in the scientific literature, such as the Lyapunov stability, stability for probability, stability in mean, practical stability. In the case of autonomous system $dx(t)/dt = Ax(t)$ there is the necessary and sufficient condition of its stability. A system is stable, if all bands of its transfer function possess the negative real part ($A < 0$).

One of the most important system properties is the property of invariance. It is a possibility to keep some desired output state for different (or determined) input actions. Let us make observations for the following system

$$y(t) = H(t)x(t) + S(t). \tag{5.13}$$

In this case the system is invariant if any (or only determined) input action $v(t)$ does not lead to undesirable responses $y(t) \pm \Delta y$.

For example, the communication systems with broadband signals (code division multiple access, CDMA) are invariant to influence of narrowband interferences of any nature possessed by restricted power. It permits combining CDMA-systems in the group signal in the same frequency band. Next they are separated using individual broadband code.

A telecommunication system is invariant to overloading if it possesses a mechanism preventing this overloading. It could be such mechanisms as RED (random early detection), WRED (weighted random early detection), SPD (separate packet dropping), TS (traffic smoothing) and a few other. A telecommunication system is invariant to failures of its components if the control mechanism is implemented providing automatic reservation (backup) of failure devices.

Therefore, the invariance of TCS can be provided only if there are necessary resources and corresponding level of service. In this case a system can execute its major functions despite of any disturbing factors and for any traffic.

5.4 Criteria of Efficiency in Control Tasks

There are some efficiency criteria which are used for obtaining optimum estimates. One of them is the criterion of minimum mean-square deviation of estimate from the true value. The following equation represents the MMSD criterion

$$J = M\left\{(x-\hat{x})^2\right\} = M(x^2) - 2\hat{x}M(x) + \hat{x}^2. \tag{5.14}$$

To find the extremum of the function y, it is necessary to find its derivative for variable x and to make it equal to zero. In the case of function (5.14) there is $2M(x) - 2\hat{x} = 0$, it gives the equality $\hat{x} = M(x)$.

In this case there is $\hat{x} = x$, it means that the estimate $\hat{x}$ is converged to the true value x. Therefore, the usage of MMSD criterion leads to the estimate without any displacement. This criterion is useful in the case of the conditions of the decomposition theorem, too.

Let us consider other optimality criteria. One of them is the maximum likelihood criterion (MLC). It is often coincides with the maximum probability criterion (MPC). Let us presume that the probability density function for deviations $(x-\hat{x})$ obeys to the normal law. In this case the MLC criterion coincides with the MMSD criterion. The normal law of density function for probabilities is shown in Fig. 5.5. This law can be represented by the following analytical expression:

$$p(x) = \frac{1}{\sqrt{2\pi\sigma^2}} e^{-\frac{(x-\hat{x})^2}{2\sigma^2}}. \tag{5.15}$$

The maximum of probability function $p(x)$ can be found in a trivial way. First of all, the derivative of the function should be found. Next, it should be set equal to zero. To simplify our task, let us take the Napierian (natural) logarithm of the function. It does not change the place of maximum because the function $\ln z$ is monotonic. It gives us the following expression for the initial function:

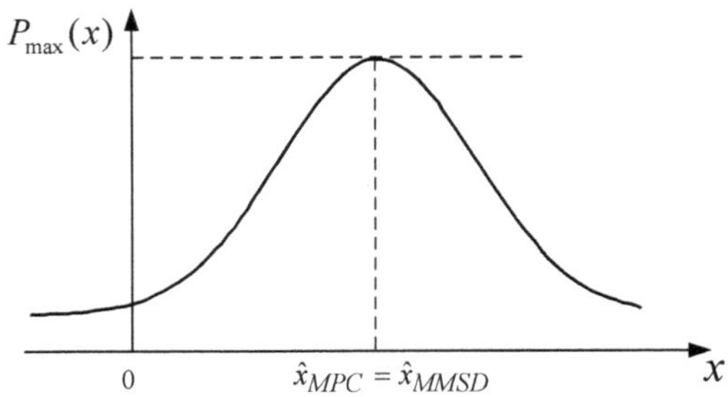

Fig. 5.5 Explanation for finding maximum probabilistic estimate

$$\ln e^{-z} = z \to \left(\frac{(x-\hat{x})^2}{2\sigma^2}\right)' \to \hat{x} = x.$$

Therefore, all discussed criteria (MMSD, MPC, and MLC) lead to the same unbiased estimates in the cases of symmetrical single-mode distributions $p(x)$. In the same time, it is not true for either multi-mode or unsymmetrical distributions (Fig.5.6). The distributions of signal phase or amplitude belong to this class, where values of $\hat{x}_{cp}$ and MPC or MLC estimates do not coincide with the true value x.

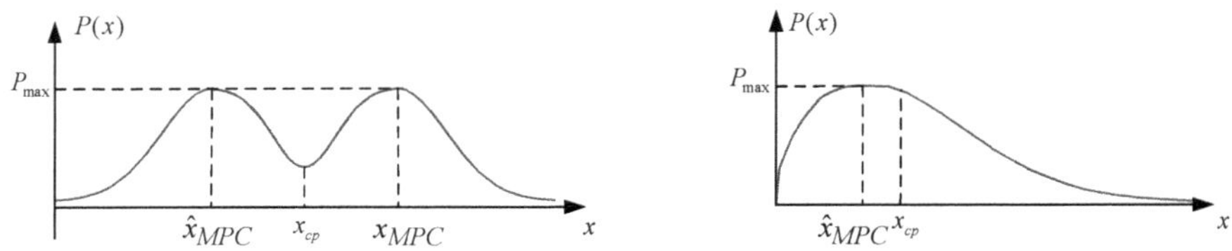

Fig. 5.6 Explanation for finding maximum probabilistic estimates for multi-mode and unsymmetrical distributions

The problem of choice of necessary criteria for given distribution is very difficult. It is solved, as a rule, on the base of engineering intuition and the context of the real task to be solved. The MMSD criterion is chosen often because it possesses some useful properties. It fines weakly for small deviations, whereas it fines strongly for large ones. It coincides perfectly well with those tasks, where it is necessary to maximize or minimize the power. The requirements of the decomposition theorem also determine necessity for use of MMSD criterion. The MLC and MPC criteria give practically similar results. They are useful if there are known the densities of distributions. Moreover, they provide rational results for single-mode probabilistic distributions.

Recommended Literature

1. Sage, A., Melsa, J.: System Identification. Academic Press, London (1971)
2. Sage, A., Melsa, J.: Estimation Theory with Application to Communication and Control. Springer, Heidelberg (1972)
3. Singh, M., Titli, A.: Systems: Decomposition, Optimization and Control. Pergamon Press, Oxford (1978)

Chapter 6
Recursive Estimates of States of Network Elements

Abstract. The chapter deals with methods used for synthesis of algorithms of recursive estimations, such as the procedure of Kalman–Busy, Robbins–Monro and so on. These algorithms are implemented in both analogue and digital variants. The peculiarities of recursive calculations of estimations are analyzed. The recommendations are given for providing stable operation modes of functioning for estimation procedures. There are given results of investigations of sensibility of Kalman-Busy filter to deviations of chosen model from the real situation.

6.1 Peculiarity of Recursive Estimations

Implicitly, a control procedure is a dynamic algorithm. It transfers some system $S(x,u,t)$ from a state x_0 into a required state x_F, satisfying to an objective (criterion) of a task to be solved. The task of estimation of state $\hat{x}(t)$ accompanies the control task. It is an important component of the control procedure and it is dynamic in the time and, probably, in the space.

Dynamic procedures are simulated by differential or difference (residual) equations. It determines their recursive nature, namely there is $x(k+1) = F(x(k))$, where k is the step of digitalization. This step can be executed either in the time ($k_t = t_n - t_{n-1}$) or in the space ($k_z = z_n - z_{n-1}$). Let us point out that the system $S(x,u,t)$ can be recursive and multi-dimensional either in time or in space or in both. The spatial recursiveness simulates the distributed nature of a system, such as the telecommunication system. Obviously, the tensor can be treated as the adequate model of such a system. But the scientific area connected with tensors is not investigated enough to apply tensors in practical applications. Let us discuss some applications connected with the time recursiveness, which are introduced now in telecommunication technologies.

The essence of recursive procedures is the following one. The value $x(k)$ obtained for the procedure step number k is used to calculate the next value for the step number $(k+1)$. Next, the value for the step $(k+2)$ is determined using the value from the step number $(k+1)$. Therefore, the consecutive values are determined

V. Popovskij et al.: Control and Adaptation in Telecommunication Systems, LNEE 94, pp. 67–82, 2011.
springerlink.com

using both previous ones and the current observation results. In other words, the recursive procedure is a calculation of the conditional mean $\hat{x}(t)/y(t)$, where $y(t)$ is a result of observation.

The first peculiarity of recursive procedures is their "short memory". They "forget" remote and even near past. Let us represent the general view of a recursive procedure:

$$\hat{x}(k+1) = \hat{x}(k) + \gamma^{-1}\Delta y(k+1,k). \qquad (6.1)$$

In (6.1) the symbol γ^{-1} stands for the scale weight ratio (coefficient), as a rule, there is $\gamma \leq 1$. The part $\Delta y(k+1,k) = y(k+1) - \hat{x}(k)$is an addition to the estimate $\hat{x}(k)$from the previous step (or a misalignment) taking into account the alterations of the current observation $y(k)$.

It follows from the equation (6.1) that the misalignment $\Delta y(k+1,k)$ to the value of estimate for the step $(k+1)$ makes correction of previous observation for a given step of digitalization. The coefficient γ^{-1} sets the degree of influence of the misalignment Δy on the convergence rate for the estimate $\hat{x}(k)$ to the required stable state. If the value of γ is near 1, then huge errors can be in the stable state. If there are small values of γ ($\gamma < 10^{-n}$, n=1,2,...,10), then the convergence rate is diminished, the memory is enlarged, but the precision is increased.

The second peculiarity makes difference between the procedure (6.1) and the mean sample estimate $\hat{x} = n^{-1}\sum_{i=1}^{n} x_i$. In the later case we can get the specific number, namely the assembly average. In the case of recursive procedure, the value of recursive estimate can be changed and corrected during the observation process. Due to correction, any new trends in the progress of the process $x(t)$ will be reflected by recursive estimates. It follows from the fact that the current estimate is corrected according with results of new observations.

The third peculiarity is in the duration of calculations. The procedure continues till the observations results $y(k)$ enter in the real time. This mode of estimation can be terminated in any instant of time, as well as it can be used for any digitalization step.

The Markov's processes possess the property of recursiveness. It follows from the equation shown below:

$$x(k) = p(x_k/x_{k-1})x(k-1). \qquad (6.2)$$

In (6.2) the probability$p(x_k/x_{k-1})$ is a probability of transition from the state $x(k-1)$ into the state $x(k)$.

It is a very important property of Markov's processes providing the following possibilities:

1. It is possible to formalize a mathematical model of the process as the state equation of the continuous process. It leads to the following equation:

$$dx(t)/dt = Ax(t) + B\xi(t). \qquad (6.3)$$

In the case of a discrete process, there is the following equation:

$$x(k) = \Phi(k,k-1)x(k-1) + B(k)\xi(k). \tag{6.4}$$

2. It is possible to find a recursive estimate using as a probability the following predictive function:

$$p(x_k/x_{k-1}) = e^{-\alpha\Delta t}. \tag{6.5}$$

In (6.5) the symbol $\alpha = A$ is a state coefficient from the expression (6.3), whereas Δt is a value of digitalization step for an interval used for making prediction.

6.2 Formalized Estimation Procedure for Random Process

This procedure can be represented in either analogue or discrete (digital) form. The kind of representation depends on a process $x(t)$ to be estimated. In both cases, three equations are used and it is enough to synthesize an algorithm for estimation a given process. These equations are the following ones: observation, state and estimation in the classical Kalman-Busy filter (KBF). They are linear equations having the closed species. But in many practical cases, either the observation (watch) equation or state equation can be nonlinear, including trigonometric functions, squares of states, and other nonlinear expressions. If a process includes nonlinear components, then the estimation task is more difficult. In these cases, some procedures are used for simplifying and linearization. There is a theory of Markov's nonlinear filtration devoted to synthesis of nonlinear algorithms. Let us point out that all practical tasks arising in TCS can be reduced to linear procedures.

In the case of an analogue process $x(t)$ the analogue KBF are used. These processes are described by the following triad of equations. The first of them is the watch equation having the following form:

$$y(t) = H(t)x(t) + v(t). \tag{6.6}$$

The second one is the state equation represented as the following:

$$dx(t)/dt = A(t)x(t) + C(t)\xi(t). \tag{6.7}$$

The estimation equation satisfying to MMSD criterion is represented in the following form:

$$d\hat{x}(t)/dt = A(t)\hat{x}(t) + V(t)H^T(t)N_v^{-1}\left[H(t)\hat{x}(t) - y(t)\right]. \tag{6.8}$$

In (6.8) the following differential equation is present:

$$dV(t)/dt = A(t)V(t) + V(t)A^T(t) - V(t)H^T(t)N_v^{-1}H(t)V(t) + C^T(t)N_\xi(t)C(t). \tag{6.9}$$

The equation (6.9) is a differential equation of Riccati for a posteriori dispersion of estimation error.

The variables N_ξ, N_ν represent spectral concentrations for power of oscillator noise in the state model and observation noise in the watch equation respectively. The first noise (N_ξ) possesses a virtual nature and it determines the degree of the process $x(t)$. The second noise (N_ν) reflects the real noise in the observation channel.

The equations (6.6) – (6.9) are represented as multi-dimensional functions with matrix coefficients, they include transposed (T) and inverse matrices having the degree -1.

Comparison of equations (6.7) and (6.8) shows that their left parts coincide, as well as the first summands from the right parts of equations. The second summand of the right part closed in square brackets is named a misalignment, it is equal to $H(t)\hat{x}(t) - y(t) = \kappa(t)$. Obviously, if the estimate $\hat{x}(t)$ coincides with the estimated random process $x(t)$ ($\hat{x}(t) = x(t)$), then the misalignment $\kappa(t)$ is close to zero. It occurs for small values of the observation $\nu(t)$), and in this case there is no need in correction for the estimate $\hat{x}(t)$. If there is a deviation from the estimated variate ($\Delta x = \hat{x}(t) - x(t)$), then the value of misalignment increases and there is $|\kappa(t)| > 0$. This inequality shows the necessity for correction to get a new estimate.

As follows from equation (6.8), the misalignment is multiplying by the value, which is opposite to the spectral concentration of the observation noise N_ν. It is clear, for large levels of the noise N_ν the quantity κ (spectral concentration of its power) is proportional to both deviation of estimate from the true value and the observation noise. Because of it, the multiplying by the opposite value N_ν^{-1} (in the case of the large observation noise) decreases confidence in the misalignment. It leads to decrease for influence of the second summand in equation (6.8). So, we can state the following: if there is $\nu \to 0$, then the main contribution in constructing estimation gives the second summand. Otherwise, the first summand is responsible for the value of estimate.

The function $V(t)$ represents the alteration of a posteriori dispersion for estimation error. The multiplying misalignment by the function $V(t)$ plays very important role in calculating the estimate. Analysis shows that the values of the function $V(t)$ are rather large after activation of the KBF. Next, these values are gradually decreased till the some value $V(\infty)$. The value $V(\infty)$ can be calculated from the equation (6.9). To do it, we should find a solution for equation $dV(t)/dt = 0$. It gives a posteriori dispersion for the stable state of the filter. In the same time, it gives the accuracy of estimate $\hat{x}(t)$ for this stable state.

If there is $dV(t)/dt = 0$, then the ordinary square equation can be got from the equation (6.9). One of solutions for this equation gives the sought-for dispersion. Let us assume that there is $H = C = 1$, and $N_\nu = P_n$ is the noise power, whereas $N_\xi = \sigma_x^2 = P_c$ is the spectral concentration of the power for a purely random useful estimated signal ($m_x = 0$). The following equation can be got from the equation (6.9):

$$V(\infty) = 2P_c/(1 + \sqrt{1 + h^2}). \tag{6.10}$$

In (6.10) the part $h^2 = 2P_c/\alpha P_n$ is a ratio of the power of useful signal to the noise power in the band of signal receiving: $F_e f = \alpha = \tau_{cor}^{-1}$. The graph of function (6.10) is shown in Fig. 6.1. If the KBF procedure is stable, then the ratio of a posteriori dispersion to a priori dispersion ($V(\infty)/\sigma_x^2$) is always less than 1.

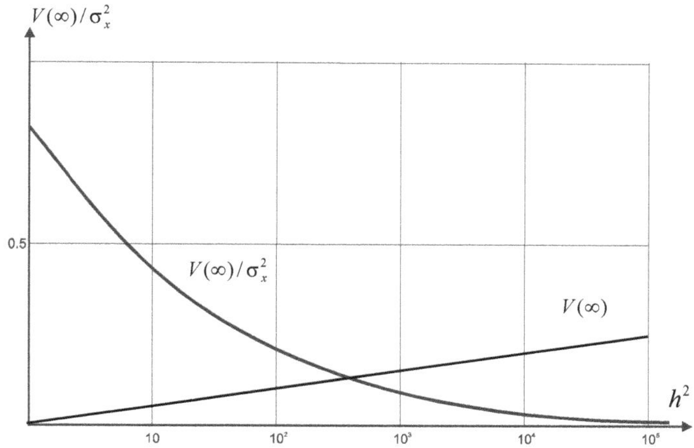

Fig. 6.1 Dependence of absolute $V(\infty)$ and relative $V(\infty)/\sigma_x^2$ a posteriori dispersion on h^2-level of signal/noise

As follows from Fig. 6.1, the value of relative a posteriori dispersion decreases steadily with increase the signal/noise ratio. It means that the relative precision of estimate increases in the same time. As follows from (6.10), the absolute value of a posteriori dispersion $V(\infty)$ increases in proportion to the estimated signal (see Fig. 6.1). It proves incorrectness of the statement that the precision of the estimate increases with the growth of the level of estimated signal $x(t)$. It is correct that the relative precision increases, but not the absolute one.

The function $V(t)$ shows the precision of estimate in KBF. It has another, very important role, namely it provides stability of an estimation procedure. Besides, it follows from (6.9) that the value of $V(t)$ does not depend on the current values of the estimated signal $x(t)$. Therefore, it is possible to calculate the precision of estimate a priori. But in practice the precision of estimate depends on precisions of parameters in the triad of equations, as well as from the choice of the quantization step Δt in discrete procedures. We discuss it a bit later.

The block diagram for observation model together with KBF is shown in Fig. 6.1.

Let us discuss this model. It includes two main blocks, namely observation and estimation blocks. The misalignment $\kappa(t)$ is generated on the output Σ_1. The derivative of estimate $d\hat{x}(t)/dt$ there is on the output Σ_2. The sought-for estimate $\hat{x}(t)$ is generated by integrator.

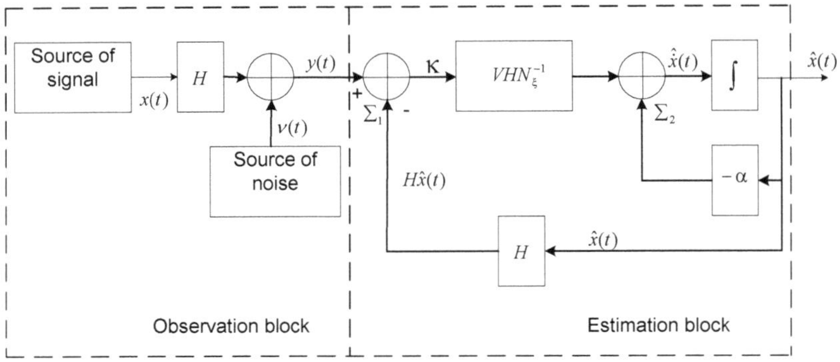

Fig. 6.2 Block diagram of Kalman-Busy filter

Analogue KBF can be easily implemented using concentrated elements of analogue electronics. They can be treated as low-pass filters or band-pass filters. Unfortunately, they are not used in practice because of inherited parameters spread.

In the case of random variates, the triad of equations (6.6), (6.7), (6.8) is simplified slightly. The watch equation remains the same ($y(t) = H(t)x(t) + v(t)$). The state equation now is presented as the following:

$$dx(t)/dt = 0. \tag{6.11}$$

The estimate equation, where there are $A = 0$, $C = 0$, now is the following:

$$d\hat{x}(t)/dt = V(t)H(t)N_\nu^{-1}\left[H(t)\hat{x}(t) - y(t)\right]. \tag{6.12}$$

In (6.12) it is presumed that there is $dV(t)/dt = V(t)H^T(t)N_\nu^{-1}H(t)V(t)$.

Equating the derivative $dV(t)/dt$ to zero, the solution $V(\infty) = 0$ is obtained. Therefore, there is the estimate $\hat{x}(\infty) = x$ for the infinite estimation interval. It means that the estimate $\hat{x}(\infty)$ convergences to the true meaning of the random variate x with zero error, that is absolutely precisely.

In the stable state there is $dV(t)/dt = 0$ and the equation (6.12) simplifies:

$$d\hat{x}(t)/dt = K[H(t)\hat{x}(t) - y(t)]. \tag{6.13}$$

In (6.13) the symbol $K = VHN_\nu^{-1}$ stands for the gain factor for the stable state. The procedure (6.13) coincides with the procedure of Robbins–Monro for stochastic approximation.

Necessity in obtaining estimates for random variates arises not only in the case when there is need in assessment of unknown constant. It arises in the case when it is necessary to estimate the mean value of some process $x(t)$, without response for its random alterations in time.

The block diagram for algorithm (6.12) is shown in Fig. 6.3. It is used for estimation of random variates. Obviously, this algorithm (estimation of random variates) is much simpler than the one shown in Fig. 6.2.

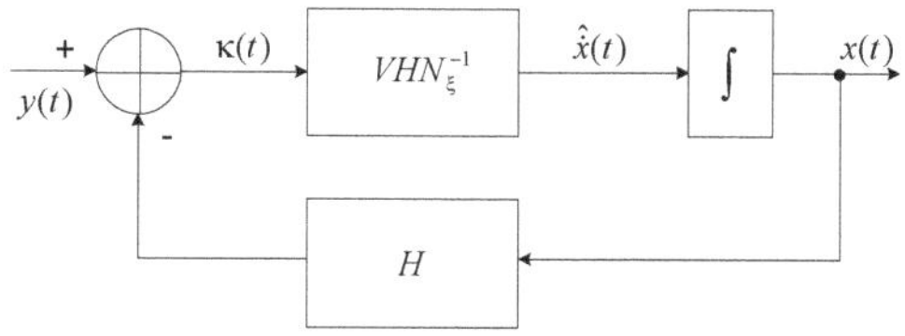

Fig. 6.3 Block diagram of analogue algorithm for estimation of random variates

Analogue algorithms are very seldom used in practice. In our book we discussed them with purely methodical goal in mind. Now, let us discuss digital algorithms implemented Kalman–Busy filters.

6.3 Digital Algorithms of Kalman–Busy Filter

The traffic inside of NGN is processed in the digital form. All future networks will preserve this tendency. Because of it, the recursive algorithms find their application only in the digital form. The discussed analogue forms of recursive algorithms are suitable for understanding the KBF theory. But the digital form is more interested, because it has many variants of practical implementation. First of all, let us show the triad of equations for digital KBF.

The watch equation is the following one:

$$y(k) = H(k)x(k) + v(k). \tag{6.14}$$

The state equation is represented in the following form:

$$x(k+1) = F(k+1,k)x(k) + G(k+1,k)\xi(k). \tag{6.15}$$

The estimate equation satisfying to MMSD criterion is shown below:

$$\hat{x}(k+1) = F(k+1,k)\hat{x}(k) + K(k)\left[H(t)F(k+1,k)\hat{x}(k) - y(k)\right]. \tag{6.16}$$

These equations include some parts which are represented as the following ones:

$$K(k) = V(k)H^T(k)N_v^{-1}; \tag{6.17}$$

$$V(k) = \left[I - K(k)H(k)\right]V(k,k-1); \tag{6.18}$$

$$V(k,k-1) = F^T(k,k-1)V(k-1)F(k,k-1) + N_\xi. \tag{6.19}$$

In these equations, the designations are kept the same as for analogue cases. The only differences represent the parts $F(k+1,k) = e^{-\alpha\Delta t}$, $G(k+1,k) = \sqrt{\sigma^2(1-e^{-\alpha\Delta t})}$, where the symbol Δt stands for the digitalization step. The block diagram of KBF is shown in Fig. 6.4.

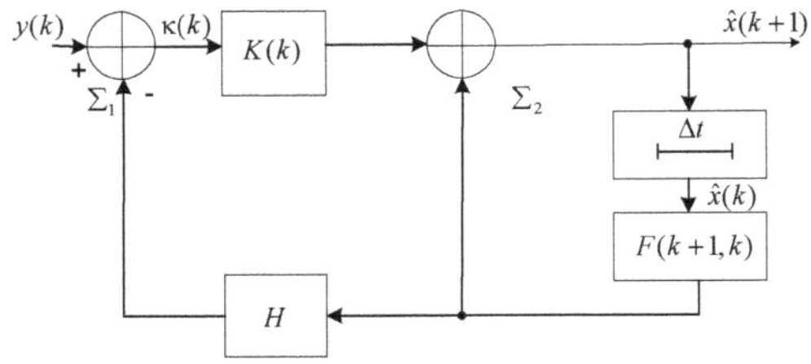

Fig. 6.4 Block diagram of digital Kalman–Busy filter

As follows from estimation equation (6.8), the misalignment is multiplied by the quantity which is reciprocal to the spectrum concentration for the power of observation noise N_V. Obviously, if there are large levels of the noise, then the value of misalignment κ shows deviation of estimate from the true value. In the same time it includes some errors because of the existence of observation noise. Multiplying by reciprocal value N_V diminish the contribution share of the second summand in equation (6.8). It can be stated that the second summand makes the main contribution in estimate if there are measurements of high quality. In this case there is $\nu(t) \to 0$. When the measurement is executed with poor quality, then the first summand makes the main contribution in the final estimate. So, everything is similar to the discussed before analogue algorithm of KBF.

Let us discuss the operation peculiarities for functioning digital Kalman-Busy filters. It is mentioned in Chapter, that it is necessary to get inter-correlative sampled values for observations of the following type: $y(k) = Hx(k) + \nu(k)$. In this case, the higher the degree of correlatedness is for samples $x(k+1)$ and $x(k)$ for all k, the higher the precision of recursive estimate is. An example of implementation of observation $y(k)$ and result of estimate $\hat{x}(k)$ are shown in Fig. 6.5. These diagrams are obtained due to mathematical simulation of KBF for the interval $T = 10\tau_{cor}$ with the digitalization step $\Delta t/\tau_{cor} = 0.01$ and for the relation signal/noise$(P_s/P_n) = 10$. The values of a posteriori dispersion $V(\infty)/\sigma_x^2$ are shown in Fig. 6.6a – Fig. 6.6c.

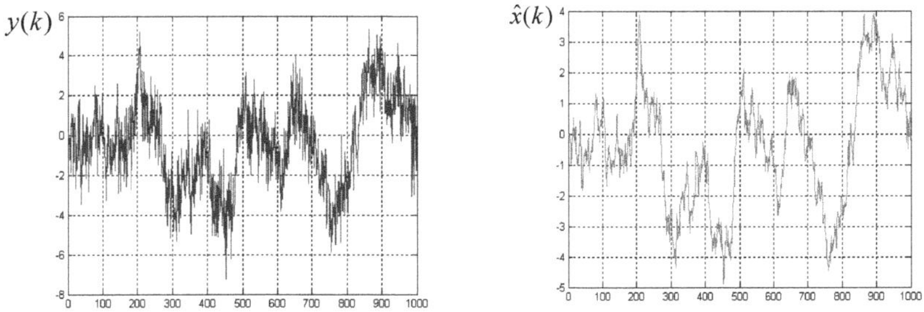

a) implementation $y(k)$ b) estimation $\hat{x}(k)$

Fig. 6.5 Implementation of observed function $y(k)$ and estimate $\hat{x}(k)$ for interval $T = 10\tau_{cor}$

As any dynamic system, the KBF possesses areas of stable and unstable functioning. The KBF operates stably, if an obtained estimate $\hat{x}(k)$ converges to the true value $x(k)$. In this case, a posteriori dispersion decreases, it becomes less than 1. After the transient period, it stabilizes on the level $V(\infty)/\sigma_x^2$, which is close to values determined from (6.10). For unstable mode (Fig. 6.6c) a posteriori dispersion $V(R)$ changes chaotically, it becomes less than 1 and never reaches the stable state.

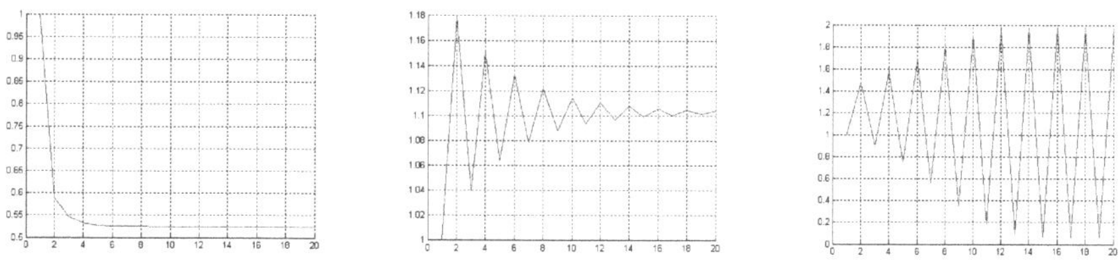

a) $T/\tau_{кор} = 0,01;\ P_c/P_ш = 10$

b) $T/\tau_{кор} = 0,01;\ P_c/P_ш = 70$

c) $T/\tau_{кор} = 0,01;\ P_c/P_ш = 100$

Fig. 6.6 Diagrams for a posteriori dispersion for error in estimation $V_x(k)$

As follows from Fig. 6.6, if only one parameter P_s/P_n is changed and other values are constant, then different convergence conditions take places for KBF. They can be either stable (Fig. 6.6a) or conditionally stable (Fig. 6.6b) and unstable (Fig. 6.6c). Let us analyze in details the modes of operations for KBF.

Starting from the first step of operation, the KBF is in its transient state, independently on the initial conditions. As a rule, the stable mode is reached for some steps. The transient state can continue from several steps to hundreds of steps (Fig. 6.6ab). It is possible that the stable state will not be reached and the chaotic process will take place (Fig. 6.6c). Therefore, the main characteristic of the transient mode is the convergence rate to the stable state.

Different factors influence the behaviour of KBF in the transient mode. The most influenced are two factors. The first of them, is the value of digitalization step $\Delta t/\tau_{cor}$ and the second is the ratio of signal/noise P_s/P_n. It is reasonable to choose as small value of digitalization step as possible, for example, to take the following values $\Delta t/\tau_{cor} \leq 10^{-2}...10^{-4}$. It is possible to achieve acceptable stable mode for bigger steps (for 10^{-1} and more). But in most practical cases the value of the step $\Delta t/\tau_{cor}$ is determined by peculiarities of this or that telecommunication technology, rather than depends on the will of a researcher. Some of these peculiarities are tremendous delays in entering either signal or control data, large time periods between adjacent messages from agents of a network, and so on. For example, the reading values for RTP protocols have the following tempo of entering: one reading per a few seconds. Taking into account that the correlation interval for traffic alterations is equal to tens of seconds, then we deal with the case when there is $\Delta t/\tau_{cor} = 0,5...0,1$. Under these conditions, it is necessary to decrease the ratio signal/noise, for example, to reach the stable mode of operation. The signal/noise ratio is a very important factor having influence on the duration of the transient mode. If there are large values of this factor, then the KBF has the slow convergence to the stable mode, sometimes it can fall into chaotic mode. The reason for such behaviour is the following one. If the level of observation noise N_v decreases, then the phenomenon of division by the small number arises in (6.17). As a result, really large numbers appear in calculations and it leads to slowing down of the computational process. To prevent this phenomenon, it is necessary to make the higher level of noise in the filter, using some artificial tools. To do it, it is possible to choose deliberately understated value of N_ξ or overstated value of N_v in the algorithm (6.17). In practice, it is useful to choose these parameters applying the method of alternate selection.

The stable mode is a main operating mode of KBF. The quality of the filter in this state is determined by the precision of estimate and the value of a posteriori dispersion $V(\infty)$. The central task for KBF programming is the choice of parameters $A = \alpha = 1/\tau_{cor}$, $\sigma_x^2 = N_\xi = P_c$, $N_v = P_n$, and $\Delta t/\tau_{cor}$.

Let us analyse the influence of errors on the precision of estimation. It could be the errors in the choice of the signal/noise ratio, as well as in the step of digitalization.

The errors in the choice of the signal/noise ratio have a very big influence. Analysis shows that the KBF is rather sensitive to deviations of chosen parameters from the parameters either of a model or of real parameters for an estimated process. It was pointed before that the errors can arise even if the P_s/P_n ratio is correct, but it is too big (more than 20...30 dB). It can result in unstable mode of filter operation. Therefore, if the P_s/P_n ratio is chosen with decreasing, it leads to insignificant errors. In the same time, the errors are really critical under increasing the P_s/P_n ratio.

The errors in the choice of the digitalization step ($\Delta t/\tau_{cor}$) have practically a one-sided nature. The decrease of this ratio is equal to increase for the number of sample readings of the estimated process for a given correlation interval. It means that decrease of the $\Delta t/\tau_{cor}$ ratio leads to increase of both estimation accuracy and stability of operation.

The following important practical conclusion can be made from this analysis. If the KBF is not stable, if it converges in a poor rate, if there are big errors of observation, then it is necessary to make a correction of a program. This correction is reduced to decrease for either the P_s/P_n ratio or for the step of digitalization $\Delta t/\tau_{cor}$.

6.4 Estimations of State for Multidimensional Systems

One-dimensional state models, corresponding solutions for estimate finding and control mode are made in the case of one element of a system. These solutions do not show such system properties as the emergence and integrity due to interrelations among the system elements. Because the properties of emergence and integrity depend on interrelations, then the necessity arises in analysis of mechanisms existed among the separate elements of a multidimensional system. Let us point out that namely these mechanisms create the integrated properties of a resulted system.

Let us use the following method for analysis of state variables. There are two mechanisms of interrelations. First of them is the interrelations through nondiagonal elements of the matrices A and C from the state equation (6.7). The second mechanism is implemented through the nondiagonal elements of the observation matrix $H$from the equation (6.6). Let us discuss these mechanisms in details.

Let us start from the multidimensional state models of the random processes.

The equation (6.7) is represented as the following system:

$$\begin{cases} \dot{x}_1 = \alpha_{11}x_1 + \alpha_{12}x_2 + \ldots + \alpha_{1n}x_n + \beta_{11}\xi_1 + \beta_{12}\xi_2 + \ldots + \beta_{1n}\xi_n, \\ \dot{x}_2 = \alpha_{21}x_1 + \alpha_{22}x_2 + \ldots + \alpha_{2n}x_n + \beta_{21}\xi_1 + \beta_{22}\xi_2 + \ldots + \beta_{2n}\xi_n, \\ \cdots\cdots\cdots\cdots\cdots\cdots\cdots\cdots\cdots\cdots\cdots\cdots \\ \dot{x}_n = \alpha_{n1}x_1 + \alpha_{n2}x_2 + \ldots + \alpha_{nn}x_n + \beta_{n1}\xi_1 + \beta_{n2}\xi_2 + \ldots + \beta_{nn}\xi_n, \end{cases} . \quad (6.20)$$

In (6.20) the variable with the sign of point above means the time differentiation (the rate of change for the function $x(t)$).

In the particular case a multidimensional system is represented by inter-independent equations, when there are $\alpha_{ij} = 0$ and $\beta_{ij} = 0$, $i \neq j$. But this assumption is true in practical cases very rarely. Moreover, the system with inter-independent states loses the system properties and, as a result, it stops being a system. Really, in TCS changing either parameters for any system element or its traffic results in necessity for correction either modes of operation or the structure of the system. It leads to redistribution of network resources, too. On the other hand, it is necessary to reflect a greater amount of system properties, then more interconnections α_{ij} and β_{ij} should be present in the model. So, the increase of the amount of interlinks leads to growth of complexness for the system model (6.20).

Obviously, it could be stated that the level of emergence and system integrity is determined by the level of interconnections among the system elements On the other hand, the growth of the amount of interconnections should not be too big. If there is the complete dependence, then the elements of such a system become indistinguishable. Obviously, there is such a level of interrelations when the integrity is the maximum one, but the system is not compressed yet to a single element. The block diagram of algorithm of double-dimensional shaping filter is shown in Fig. 6.7. Obviously, the increase of the dimensionality leads to the increase for the complexness of the structure.

It is known from the theory of linear multidimensional equations that the general form of the system (6.20) can be simplified due to nondegenerate sine-cosine transformation. Therefore, it can simplify the algorithm for forming the multidimensional process $\vec{x}(t)$. In the case of double-dimensional system of equations this transformation is reduced to applying the operator of rotating for coordinate axes by the angle γ:

$$\begin{pmatrix} \sin\gamma & \cos\gamma \\ -\cos\gamma & \sin\gamma \end{pmatrix}. \quad (6.21)$$

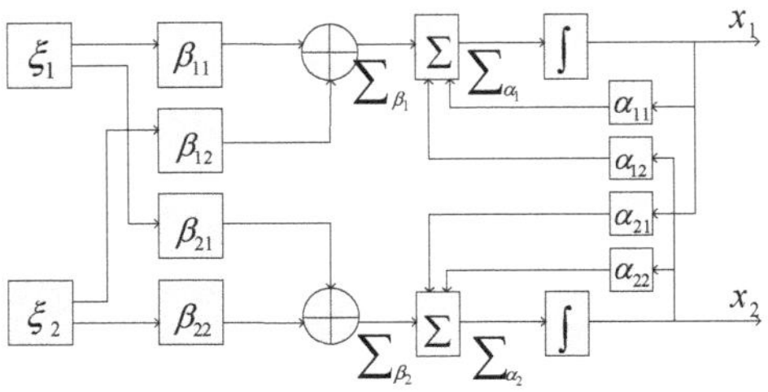

Fig. 6.7 Block diagram of double-dimensional shaping filter

Using transformation (6.21), it is possible to find such a turn of coordinates in the system x_i, $i = 1,2$, when the matrix A is diagonal under the complete matrix B. It is possible too finding the turn leading to the complete matrix A under the diagonal matrix B. Therefore, diagonalization of the matrix A results in implementing interrelations between elements x_i and x_j using elements β_{ij}. If the diagonalization of the matrix B is executed, then the interrelations are implemented using elements α_{ij}.

The stability of state model is provided by two conditions. The first of them is the requirement for all α_{ii} to be less than zero. The second condition is formulated as $\max\limits_{1 \le i \le n} \left(\sum_{j=1}^{n} \left| \alpha_{ij} \right| \right) \le 2$. This condition provides lack of increase for the state $x_i(t)$ if the value of t increases.

Let us discuss the multidimensional models for observation of random processes. Mutual connection between components x_i and x_j can be formed under measurements. For example, the measuring value x_i can add portions of other components x_j together with the main result for x_i. In this case the watch equation for the x_i is the following one: $y_i = h_{ii}x_i + h_{ij}x_j + \ldots + h_{in}x_n + v_i$.

Therefore, the matrix H in the watch equation becomes complete and the watch equations are interconnected:

$$\begin{cases} y_1 = h_{11}x_1 + h_{12}x_2 + \ldots + h_{1n}x_n + v_i, \\ y_2 = h_{21}x_1 + h_{22}x_2 + \ldots + h_{2n}x_n + v_2, \\ \cdots\cdots\cdots\cdots\cdots\cdots\cdots\cdots\cdots\cdots \\ y_n = h_{n1}x_1 + h_{n2}x_2 + \ldots + h_{nn}x_n + v_n. \end{cases} \tag{6.22}$$

Very often, these interconnections have the negative nature, because they do not permit separation for this or that process in its pure form.

The equations for observed signals are analogous to (6.22) in the wireless technologies Wi-Fi and WiMAX. In these technologies the multidimensional spatial-temporal encoding is used, which is known as multiinput/multioutput (MIMO) encoding. In the same time, existence of cross connections in the watch equation (6.22) does not add integrated properties to the system $S(x,u,t)$. Rather, these cross connections are similar to mutual interferences and they should be either suppressed or compensated.

Thus, there are three mechanisms for taking into account the mutual connections among components x_i of multidimensional system: due to state coefficients α_{ij}, due to generation coefficients β_{ij}, and due to reciprocal influences under measurements h_{ij}. Any of these mechanisms can act by itself, or it can appear together with other mechanisms. The choice of mechanism belongs to a researcher and he/she acts in dependence on the context, sense and convenience of operations in the task to be solved.

There is one peculiarity in operating discrete multidimensional Kalman-Busy filters. In this case, the matrices $F(k+1,k)$ and $G(k)$ have not only diagonal elements F_{ij} and G_{ij}, determining properties of the state i of the component x_i. They also include nondiagonal elements F_{ij} and G_{ij}, providing account of cross (mutual) correlation between components x_i and x_j.

The algorithm of multidimensional KBF (6.16) should correspond to the multidimensional state model (6.20). It can be used either the model with two complete matrices A and B, or the model with one diagonal matrix and one complete. Let us show the analytical form of two-dimensional state equation with two complete matrices of state and generation:

$$\begin{cases} x_1(k+1) = F_{11}(k+1,k)x_1(k) + F_{12}(k+1,k)x_2 + G_{11}(k)\xi_1(k) + G_{12}(k)\xi_2(k), \\ x_2(k+1) = F_{21}(k+1,k)x_1(k) + F_{22}(k+1,k)x_2 + G_{21}(k)\xi_1(k) + G_{22}(k)\xi_2(k). \end{cases} \tag{6.23}$$

Let us represent the algorithm for estimate the components of two-dimensional KBF, where the connections between the components x_1 and x_2 are executed only due to coefficients F_{12} and F_{21}. This algorithm can be represented as the following system of equations:

$$\begin{cases} \hat{x}_1(k+1) = F_{11}\hat{x}_1(k) + F_{12}\hat{x}_2(k) + K_{11}\left[H_{11}F_{11}\hat{x}_1(k) - y_1(k)\right], \\ \hat{x}_2(k+1) = F_{21}\hat{x}_1(k) + F_{22}\hat{x}_2(k) + K_{22}\left[H_{22}F_{22}\hat{x}_2(k) - y_2(k)\right]. \end{cases} \tag{6.24}$$

The block diagram of KBF algorithm corresponding to (6.24) is shown in Fig. 6.8.

It is possible to represent the block diagram of the two-dimensional KBF in some different form. Obviously, it is possible to take into account the interrelations among components using coefficients G_{12} and G_{21}. In this case, the coefficients F_{12} and $F_{21}$are deleted. If the matrices F and G are complete, then it leads to appearance of connecting elements K_{12} and K_{21}. If mutual influences are modelled only due to elements of the matrix H_{ij}, then the mutual connections into KBF are taken into account due to coefficients $K_{ij} = V_{ij}H_{ij}N_v^{-1}$.

Neglect of interrelations between the components results in the loss in the quality of estimates. It leads to diminish the control quality. Such a loss of quality results in neglecting the most important system properties, namely the properties of integrity and emergence.

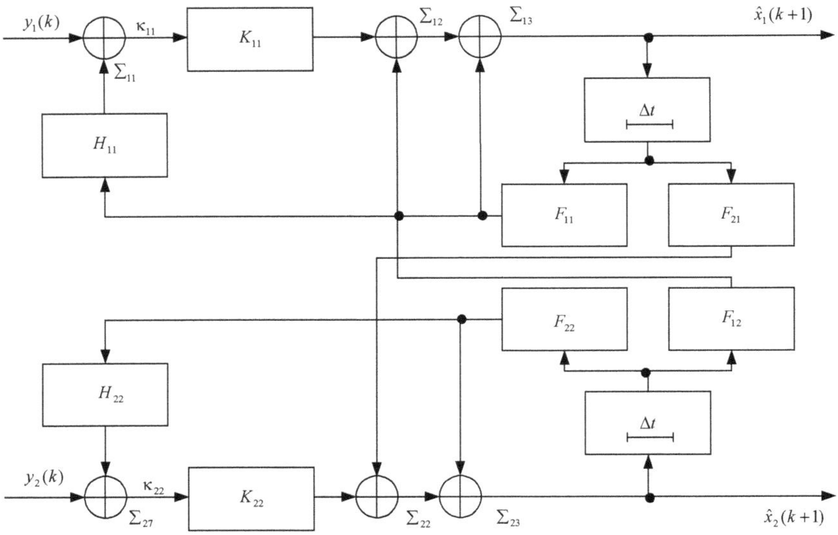

Fig. 6.8 Block diagram of two-dimensional Kalman-Busy filter

6.5 Peculiarities of Estimation of Discrete Variates

In up-to-day telecommunication systems there are used digital algorithms of recursive estimation of variates. The triad of equations should be transformed. If there is $H = 1$, then the watch equation is simplified. Now it is represented as the following one:

$$y(k) = x(k) + v(k). \tag{6.25}$$

The state equation corresponding to (6.11) is represented in the following form:

$$x(k+1) = x(k).$$

The estimate equation can be written as the following one:

$$\hat{x}(k+1) = \hat{x}(k) + K(k)\left[y(k) - \hat{x}(k)\right]. \tag{6.26}$$

In (6.26) the symbol $K(k)$ is a weighted coefficient satisfying in the general case to the following coefficient:

$$\sum_{k=1}^{n} K(k) \to \infty, \sum_{k=1}^{n} K^2(k) < \infty, 0 < K(k) \leq 1. \tag{6.27}$$

The algorithms used for recursive calculation of the mean (6.12) and (6.26) are known in mathematics as the procedures of stochastic approximation of Robbins–Monro (RM). The block diagram of algorithm (6.26) is shown in Fig. 6.9.

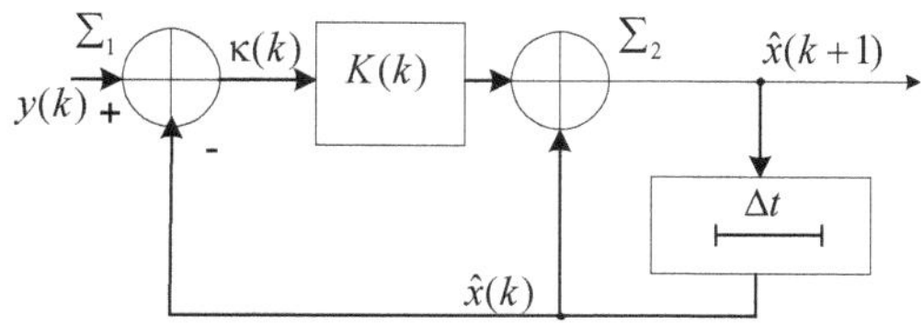

Fig. 6.9 Block diagram of algorithm for estimation of variate $x(k)$

For example, the sequence $K(k) = n^{-1}$, where $n = 1,2,3...$, satisfies to conditions (6.27). The practice shows that the procedure (6.26) is stable under other conditions, which are weaker than the restrictions (6.27). These restrictions are the following:

$$0 < K(k) = const \leq 1. \qquad (6.28)$$

In telecommunication systems, the procedure (6.26) appears in slightly different, transformed form. So, clearing brackets in equation (6.26) and regrouping variables leads to the following equation, where the condition (6.28) is taken into account:

$$\hat{x}(k+1) = \hat{x}(k) + Ky(k) - K\hat{x}(k) = \hat{x}(k)\,[1-K] + Ky(k). \qquad (6.29)$$

In (6.29) there is equality $K = 2^{-n}$, $n = 1,2,\ldots,10$.

The role of the step constant K can be explained taking into account the parabolic nature of MMSD criterion. During recursive assessment for the estimate Δx, the error is getting down along the parabola to the stable meaning. Each step of this descent is accompanied by decrease of the error.

Three situations are shown in Fig. 6.10 explaining the influence of the weighted coefficient $K(k)$ and value of the post-tuning drift Δx. If the value of $K(k)$ is optimal, then the procedure converges rapidly and the post-tuning drift is very small ($\Delta x \to 0$). This situation is shown in Fig. 6.10a. If $K - const$ is unchangeable and there is a big step (Fig. 6.10b), then the procedure converges rapidly but there is rather big value of the post-tuning drift Δx. If the value of K is small, then it be relatively large amount of steps to reach the stable state, but post-tuning drift is rather small. The last situation is shown in Fig. 6.10c.

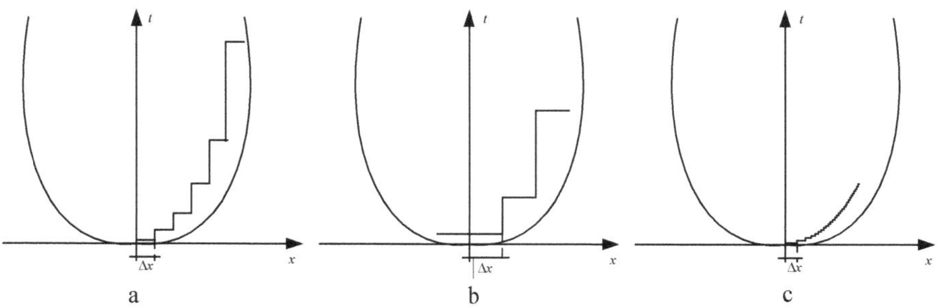

Fig. 6.10 Influence of step constant

The value of coefficient K determines the convergence rate for both procedures (6.26) and (6.29). Because of it the coefficient K is called the step constant.

The procedure (6.29) is used in RED algorithms targeted preventing overloading of routers, as well as in procedures of estimate for rotary circulation of packets (round-trip time, RTT) and some other.

The procedures (6.12), (6.26), and (6.29) are used very often by network researchers. In the case of estimation of mean the coefficient K is chosen from the interval (0,9...0,1). As analysis shows, the procedure reaches the stable mode for 20...200 digitalization steps. As a rule, the accuracy of estimate decreases with acceleration of convergence rate (when there is $K \to 1$).

Very often in practice, the RM procedures are used for estimation of random processes. But the procedure possesses the smoothing property and it leads to loss of rapid alterations of a process. This property increases with decrease of the step constant. For example, in the case of RED strategy the algorithm (6.18) is used for smoothing the rapid variations in intensity of packet flow. In this case the value of the step constant is equal to $K = 2^{-8}...2^{-9}$. The rapid changes and outliers of intensity have no influence on this estimate. But the slow component of the trend of nonstationarity is selected. In this case, the trend of nonstationarity is selected due to another property of recursive procedures, which is the lack of the procedure's memory.

Recommended Literature

1. Sage, A., Melsa, J.: Estimation Theory with Application to Communication and Control. Springer, Heidelberg (1972)
2. Singh, M., Titli, A.: Systems: Decomposition, Optimization and Control. Pergamon Press, Oxford (1978)

Chapter 7
Synthesis of Control Algorithms for Telecommunication Systems

Abstract. The chapter is devoted to synthesis of control algorithms. Two main approaches are discussed, namely, the control of the state and the control of the observation. Both approaches are considered from the point of view of the method of state variables. The examples are given for solution of the tasks of state control under fulfilment of conditions for the decomposition theorem. The problem of observation control is formulated as a task of control of the observation basis. This control should provide some necessary properties of the observed useful signal, such as its interference protection. The last class of problems is reduced to construction of adaptive compensator of interferences and adaptive antenna arrays, as well as to the tasks of spatial-temporal encoding and spatial-temporal access to the base station in the system of mobile communications.

7.1 Short Introduction into Control Methods

Different control methods are used in telecommunication systems. Among them there are situational methods, which are based on the logic of people making decisions, automatic methods used for message switching, and automated methods providing execution of simple, rather routine operations. Now, more and more procedures are executed automatically, using different optimization methods. It permits getting the maximum effect from control operations. In the same time, the requirements are increasing for observation and control channels, alarm systems, measuring devices (agents, sensors, transudes), and for regulators and execution units.

The control methods based on Ponselle's principle are used widely in existent technologies. This class of algorithms includes all algorithms of activation and shutdown based on alarm system, algorithms for distribution and control of network resources, and so on. These algorithms belong to technological issues and they are implemented, as a rule, by program centralized methods.

The control methods using Watt's principle are widely used, too. They are implemented as a response for some external influence. This class of problems includes

V. Popovskij et al.: Control and Adaptation in Telecommunication Systems, LNEE 94, pp. 83–96, 2011.
springerlink.com

tasks of mode control for individual network elements (amplification, synchronization, and stabilization), control of network characteristics (the turn-around time, preventing overloading of a router, and so on). These methods are implemented in both centralized and decentralized variants. Their implementation is based on the methods of theory of optimal control, using state variables. Two main control types are considered in the methods of state variables: the control of the system state and the watch (observation) control. These types of control lead to different algorithmic solutions. Let us discuss the control algorithms in details.

7.2 Control of System State

First of all let us discuss the general formulation of the task of optimal control. This type of control targets a transition of the system state from some phase coordinates into other to reach either required structure or mode of some network element (or of the total network). To find the required control, some optimality criterion is chosen. As a rule, the mean-square criterion is chosen:

$$J(\mathbf{x},u) = \frac{1}{2}x^T(t_F)Dx(t_F) + \frac{1}{2}\int_0^{t_F}\left[x^T(t)Qx(t) + u^T(t)Ru(t)\right]dt. \tag{7.1}$$

In (7.1) the symbol t_F is a final time or the time of control goal achievement.

If there is a stochastic system, then the criterion (7.1) is replaced by the mathematical expectation of criterion:

$$M\{J(\bar{x},y)\} \to \min_x . \tag{7.2}$$

The value of control $u(t)$ is finding by substitution of some state equation into either (7.1) or (7.2). For deterministic system the state equation is the following one:

$$dx(t)/dt = Ax(t) + Bu(t). \tag{7.3}$$

In the case of stochastic system the state equation is represented in the following form:

$$dx(t)/dt = Ax(t) + Bu(t) + C\xi(t). \tag{7.4}$$

The telecommunication systems operate with a lot of random processes, such as random traffic, random signals, and random noises. Because of it, it is necessary to use the criterion (7.2). In this case the conditions of decomposition theorem allow replacement the variable $x(t)$ by the value of estimate $\hat{x}(t)$ into equations (7.1)-(7.4). Next, the control can be executed using the deterministic approach. The block diagram is shown in Fig. 7.1 representing the system controlled by state.

In accordance with the control theory, the optimal trajectory for transition into the required phase state is determined by minimizing Hamiltonian $\mathrm{H}(t)$ along this trajectory:

$$d\mathrm{H}(t)/du \to 0. \tag{7.5}$$

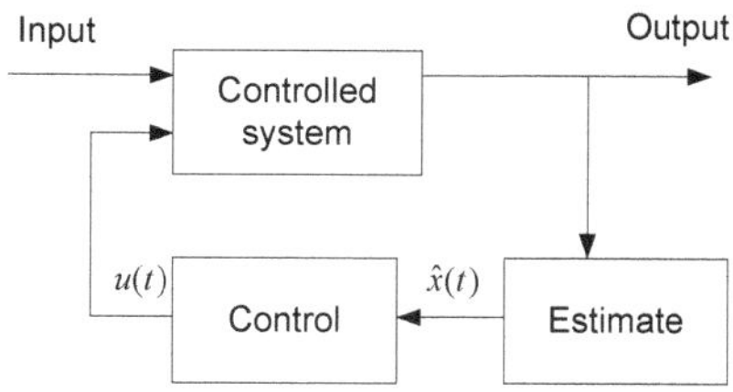

Fig. 7.1 Block diagram of controlled system with separated blocks of stochastic estimate and deterministic control

In (7.5) the equation $\mathrm{H}(t) = \frac{1}{2}x^T(t)Qx(t) + \frac{1}{2}u^T(t)Ru(t) + \lambda^T(t)Ax(t) + \lambda^T(t)Bu(t)$ is the Hamiltonian.

If condition (7.5) takes place, then the optimal control corresponding to (7.5) is determined as the following one:

$$u(t) = R^{-1}B\lambda(t). \tag{7.6}$$

In (7.6) the equation $\lambda(t) = D(t_F)x(t_F)$ determines reduced state of the controlled system, which is reached in the instance t_F.

After transformations, the following equation can be got:

$$u(t) = L(t)x(t). \tag{7.7}$$

In (7.7) there is $L(t) = -R^{-1}B^T P(t)$, $dP(t)/dt = -P^T(t)A - A^T P(t) + P^T(t)B R^{-1}BP(t) - Q$ is the Riccati equation.

In the case of a discrete system there is

$$u(k) = L(k,k-1)x(k-1). \tag{7.8}$$

In (7.8) there is $L(k,k-1) = -R^{-1}B^T P(k)$, and $P(k) = Q(k) + A^T[P^{-1}(k-1) + B R^{-1}B]A$.

The discussed control is named either terminal or final. This name is connected with the fact that the control is finished after the time t_F and the system transits into a stable state. But in many cases the control is used for either monitoring or supporting this or that state for a considerable time. In this case the upper limit of integration in (7.1) is equal to infinity.

As follows from this discussion, the internal content of a system is changed under the control actions. This content includes either structure or function of the controlled system. The causes of changes can be either external (change of traffic) or internal (failure of some components). Let us discuss an example of solution of the task with state control.

Let us discuss the task of control for the state of a network element. The typical case is the case of control for infinite time interval. It means that the task is solved for supporting of some required state. In the case of MMSD criterion, let us use the expression (7.1) without its first summand, namely:

$$J = \int_0^\infty (x^2(t)Q + u^2(t)R)dt, \tag{7.9}$$

In (7.9) the following state equation of controlled stochastic system presents:

$$dx(t)/dt = Ax(t) + Bu(t) = -\alpha x(t) + 2\alpha\sigma_x^2\xi(t) + Bu(t). \tag{7.10}$$

In many practical cases the costs of control are not principal for control tasks. It permits further simplification of the equation (7.9), namely the following expression $J = \int_0^\infty x^2(t)Qdt$ can be obtained.

Let us presume that the state $x(t)$ is random. In this case the approach (7.2) should be used for criterion (7.9), but it is connected with necessity in use special methods for integration of stochastic functions. It is more reasonable to apply the conditions of decomposition theorem. It permits using the estimate $\hat{x}(t)$ instead of the random process $x(t)$. The control (7.6) is finding in the following form:

$$u(t) = -\frac{BP(t)\hat{x}(t)}{R}. \tag{7.11}$$

The represented expression (7.11) used for control includes both transient and stable modes of the element. For the stable mode, the control dynamic depends only on the state estimate

$$u(t) = -\frac{BP}{R}\hat{x}(t). \tag{7.12}$$

It is taken into account in (7.12) that there is $P(t) = P$ for $dP(t)/dt = 0$ in the stable state.

Taking into account the state 7.10, the estimate $\hat{x}(t)$ can be found from the following equation

$$d\hat{x}(t)/dt = -\alpha\hat{x}(t) + K(y(t) - \hat{x}(t)). \tag{7.13}$$

In (7.13) the equation $y(t) = x(t) + v(t)$ is a watch equation, whereas the symbol $K = P/N_v$ stands for amplification coefficient of KBF (7.13), where the symbol N_v determines the spectral density of observation noise power $v(t)$.

In the same manner the control for discrete systems can be found. The state equation for such system is the following one:

$$x(k+1) = e^{-\alpha\Delta t}x(k) + \sqrt{\sigma_x^2(1 - e^{-\alpha\Delta t})}\xi(k) + Bu(k). \tag{7.14}$$

In (7.14) there is $\alpha = \tau_{cor}^{-1}$ and Δt is the digitalization step.

The control corresponding to (7.12) can be expressed as the following one:

$$u(k) = -\frac{BP}{R}\hat{x}(k). \tag{7.15}$$

To find the estimate for a discrete state of the system (7.14) the following equation can be used:

$$\hat{x}(k+1) = e^{-\alpha\Delta t}\hat{x}(k) + \frac{P}{N_v}(y(k) - \hat{x}(k)). \tag{7.16}$$

The accuracy of control is determined by the accuracy of state estimate, because there are no other unknown parameters in (7.15). In practice, it is quite possible that the chosen model (either (7.10) or (7.14)) has some deviations from the real state. It can result in some additional losses. Investigation of quality losses connected with deviations of the model is a very important independent problem. It is discussed in Chapter 8 of our book.

7.3 Control of System Observation

In contrast with the state control, the observation control does not assume any changes in the internal properties of a controlled system. On the contrary, some changing in the observation base can be done due to control procedures. These changing can be the following ones: alterations of the antenna's orientation, providing better coordination with an observed object, suppressing of interferences, and so on. Obviously, transformation of single-dimensional basis cannot lead to any alterations. Alterations are possible if the number of dimensions is more than two (dim > 2). It is necessary to have at least two-dimensional base to obtain the desired conditions for observation of useful signals. It can be done by orthogonalization of the basis in respect to either interference or useful signal. Either spatial or polarization orts can be used as base functions, as well as either temporal or frequency functions. For example, it is always possible to find such a direction for a two-dimensional antenna where the level of received interference tents to zero ($P_i \to 0$). The choice of base is made in terms of the task to be solved, as well as availability of this or that resource.

The purpose of this control is achievement the desired ratio of signals and interferences on the system input. In the case of state control, the system is transited from some point of the phase space into another point. In the case of observation control, the desired observation is formed due to change of the scale and shift of some base functions. It results in the weighting of observation components.

Let us explain the importance of observation problem. It is very important to make adequate representation of received information under measurement and observation processes. It means that this information should be represented without interferences, in some particular coordinates, in some determined scale, in some desired angle and so on. It means that the observation control is reduced to finding the most appropriate conditions for observation (or measurement). Let us point out that in our everyday life we act in the same manner. For example, in a sunny day we narrow our eyes when we are trying to discern some object.

The mutual interferences appear because of the high utilization of the radio-frequency spectrum. In this case they say about violation of conditions for electromagnetic compatibility (EMC). Very often, it is impossible to avoid interferences using either change of the frequency or applying some other methods. Different methods are used for suppressing of interferences. For example, they use either the signals with pseudorandom alteration of operating frequency or the wideband signals in the technology IEEE 802.15. The usage of these signals can be viewed as an invariant approach for the struggle with interferences. But it requires tremendous throughput of communication channels. The methods of control for observation base are more economical, they are reduced to the tasks of interference suppressing (rejection or compensation).

Three main methods are known for solution of tasks connected with improving of interference protection. Let us briefly outline them.

1. If there is some interference for observation, it should be compensated. Such an approach is named the adaptive compensation of interferences. This approach is used, for example, for improving electromagnetic compatibility in medicine observation. For example, task should be solved when it is necessary to measure the pulse rate of a child in the mother womb on the background noise of the mother pulse.

2. Let it be some standard y_e of a desired useful signal. In this case the task is reduced to follow this standard. The misalignment $\kappa(t) = (y(t) - y_s)$ is used as a control signal. This principle is used in adaptive watch systems, adaptive regulative devices, and adaptive antenna arrays.

3. The methods of spatial-temporal encoding are used in radio channels with multipath propagation. It permits increasing throughput of radio communication line up to 1,6 – 1,8 times for a dedicated frequency.

Let us discuss the typical situation taking place in ordinary communication channel of cellular communication (or any another wireless system). Let the watch equation include three components: the useful signal $c(t)$, the interference $n(t)$, and the white noise $\nu(t)$. It gives the following equation:

$$y(t) = c(t) + n(t) + \nu(t). \tag{7.17}$$

The spectra of useful signal and interference can be overlapped. The typical situation of overlapping is shown in Fig. 7.2.

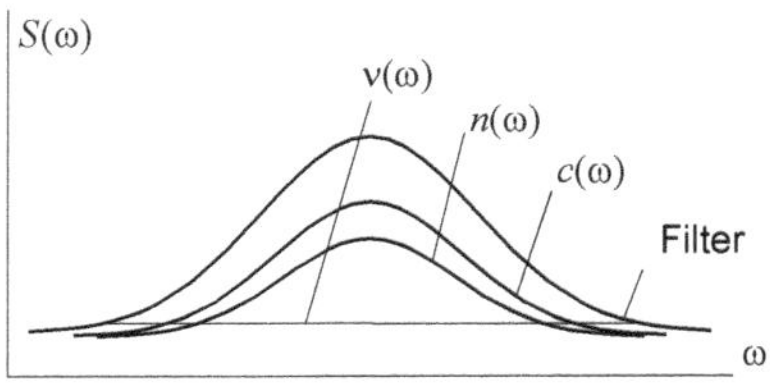

Fig. 7.2 Relation for spectra of filter, signal $c(\omega)$, interference $n(\omega)$, and noise $\nu(\omega)$

As follows from Fig. 7.2, it is impossible to obtain satisfactory quality of receiving of signal $c(t)$ if the levels of useful signal P_s and interference P_i are commensurable. In this case it is more preferable to use the compensation of interference $n(t)$. This approach can increase the quality of received signal despite of interferences and existence of the white noise.

7.4 Control of Adaptive Interference Compensators

Let us discuss the situation when the spectra of the signal $c(t)$ and the interference $n(t)$ coincide partially or completely. In this case the following two approaches can be applied. The first of them is reduced to the replacement of observation channel. For example, it could be done by changing the frequency, or using noise combating codes, or some other changes. But such mode of action is not always possible due to already used configuration or because of the time losses connected with these actions. The second approach is reduced to giving a system a special ability for compensation of interferences. It is executed by the special adaptive interference compensator (AIC). Let us discuss the problem of interference compensation.

Let it be the watch equation (7.17) including signal $c(t)$, interference $n(t)$, and noise $v(t)$. It is necessary to form such a signal (or an anti-interference) $n^*(t) \approx -n(t)$, which could be subtracted from the observed implementation. It gives the following equation:

$$y^*(t) = c(t) + n(t) - n^*(t) + v(t) = c(t) + v(t) + \Delta n(t). \qquad (7.18)$$

In (7.18) the symbol $\Delta n(t)$ stands for the remainder of uncompensated interference which should be minimized($\Delta n(t) \rightarrow 0$).

But it is possible that the level P_s of the useful signal $c(t)$ is rather large in the main observation channel (see Fig. 7.2). In this case, it prevents forming required anti-interference $n^*(t)$. It results in the absence of the direct solution. In this connection, the problem of compensation is expanded and it is solved in two stages. The first stage is finding possibility for creating separate auxiliary reference channel to suppress the useful signal $c(t)$. The second stage is connected with formation of anti-interference $n^*(t)$, which helps in final compensation of the interference in the main receive channel. Let us discuss a way for solution of this task.

The essence of the task is in formation of anti-interference $n^*(t) = -n(t)$, where $n^*(t)$ is an interference equal by amplitude to the interference $n(t)$existing in the main receive channel and opposite in the phase. To solve this problem, it is necessary to create the reference channel represented by the following equation:

$$y_{ref}(t) = n_{ref}(t) + v_{ref}(t). \qquad (7.19)$$

In (7.19) the symbol $n_{ref}(t)$ stands for interference, which is the same as the interference in the main channel but it could have different amplitude and phase. The reference channel $y_{ref}(t)$is the second observation channel. It means that these two channels form double-dimensional basis, where the conditions (7.9) and (7.19) take places. The second channel (7.19) can be created by the help of second auxiliary antenna, for example, or in some other way. Let us point out that the task of creation of this channel is very difficult and its solution requires real engineering art.

When the reference channel $y_{ref}(t) = n_{ref}(t) + v_{ref}(t)$ is ready, then the further task is reduced to finding such a complex weighted coefficient $\dot{W} = |W| e^{-j\varphi}$ that can be multiplied by components of the reference channel and gives us the following result:

$$\dot{W} y_{ref}(t) = W n_{ref}(t) + W \nu_{ref}(t). \tag{7.20}$$

In (7.20) the symbol $W y_{ref}(t)$ denotes the required anti-interference. Next, the signals from both the main channel (7.18) and the reference channel (7.20) enter in antiphase the common adder Σ_1 (Fig. 7.3).

Therefore, according to (7.18) the following result is obtained as the output of the common adder:

$$y^*(t) = c(t) + n(t) - \dot{W} n_{ref}(t) + W \nu_{ref}(t) + \nu(t). \tag{7.21}$$

In (7.21) there are $n(t) - \dot{W} n_{ref}(t) = \Delta \text{n(t)} \to 0$ and $W \nu_{ref}(t) + \nu(t) = \nu^*(t)$is treated as a Gaussian white noise.

To solve the task of estimation for the weighted coefficient W let us use the formalization of Kalman-Busy filter. It means that we can use the following equation:

$$\hat{x}(t) = A(t)\hat{x}(t) + VN^{-1}H\left[y(t) - H\hat{x}(t)\right]. \tag{7.22}$$

Let us assume that $x(t)$ is the same as the complex weighted coefficient W. In this case the misalignment is represented as $\kappa(t) = y(t) - y_{ref}W(t)$. Thus, there were the following replacements in the KBF procedure: x is replaced by W, and H is replaced by $y_{ref}(t)$. As a result, the following algorithm for estimation of the optimal value of $W(t)$ can be formulated:

$$d\hat{W}(t)/dt = A(t)\hat{W}(t) + K(t)y_{ref}(t)\left[y(t) - \hat{W}(t)y_{ref}(t)\right]. \tag{7.23}$$

In (7.23) the coefficient $K(t) = VN^{-1}$ is a constant in the stable mode.

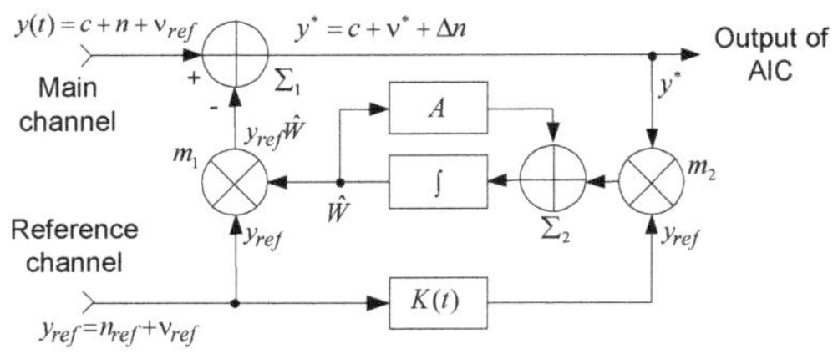

Fig. 7.3 Block diagram of adaptive interference compensator

The algorithm (7.23) coincides in some details with the known algorithm of Widrow. The Widrow's algorithm is represented as the following one:

$$d\hat{W}(t)/dt = K(t)\left[y(t) - \hat{W}(t)y_{ref}(t)\right]y_{ref}(t). \tag{7.24}$$

Comparison of Widrow's algorithm (7.24) and algorithm of Robbins–Monro (6.13) used for estimation of variates leads us to the following conclusion: these algorithms are similar in both form and function. The generalization of Robbins–Monro algorithm leads to Kalman–Busy filter. Obviously, the Widrow's algorithm can be reduced to KBF. Therefore, the algorithm (7.24) is optimal for the case $dW(t)/dt = 0$. It is the case when parameters of interference $n(t)$ are not changed in time. Of course, this interference is a variate.

The discrete AIC algorithm based on the RM procedure is represented as the following one:

$$W(k+1) = W(k) + K(k)\left[y(k) - Wy_{ref}(k)\right] y_{ref}(k). \tag{7.25}$$

In (7.25) the coefficient $K(k)$ can be constant, as it is for the Widrow's algorithm.

The structure of algorithm (7.25) coincides completely with the known discrete Widrow's algorithm. Here we can see the direct connection with the estimation algorithm for variates. The block diagram of AIC Widrow's algorithm based on (7.24) is shown in Fig. 7.3.

The discrete algorithm of AIC based on KBF is the following one:

$$\hat{W}(k+1) = \Phi(k+1,k)\hat{W}(k) + K(k)\left[y(k) - \hat{W}(k)y_{ref}(k)\right] y_{ref}(k).$$

Obviously, the KBF algorithm operates optimally if the interference $n(k)$ has a random change of its spatial spectrum.

Let us make interpretation for operation of AIC. The interference y_{ref} from the reference channel weighted by the coefficient $\hat{W}$ enters the main channel with the negative sign. It leads to subtraction using the adder Σ_1. The adder Σ_1 produces result $y^*(t)$. Next it should be multiplied by y_{ref} using the unit of multiplication m_2. The output of m_2 contains the following result:

$$\begin{aligned} y^*(t)\cdot y_{ref} &= (c + v^* + \Delta n)(n_{ref} + v_{ref}) = \\ &= cn_{ref} + cv_{ref} + v^* n_{ref} + v^* v_{ref} + \Delta n v_{ref} + \Delta n n_{ref} \end{aligned}. \tag{7.26}$$

After integration, all summands (except the last one) are in average equal to zero. It is true because these summands are not inter-correlative. The sixth summand is not equal to zero:

$$u(t) = \int n\Delta n dt \neq 0. \tag{7.27}$$

The expression (7.27) follows from the fact that both n and Δn are correlative, moreover, they are the copies of each other. Thus, we can get the control signal $u(t) = \int n\Delta n dt$. This signal enters the multiplication device m_t and it is active till the remainder of interference Δn reaches its minimum value: $\int n\Delta n d(t) \to 0$. It leads to compensation of the interference $n(t)$ in the main channel.

There are many methods used for creating the auxiliary reference channel. All these methods include the procedure for deleting useful signal. Let us discuss three of them.

1. Pauses are found or created in the structure of useful signal $c(t)$ for creation of the reference channel. The adapted interference compensator operates in the time of test-pauses or test-signals (they are known signals for the case when there is $y_{ref} = u + v$. Some available service signals can be used as the test signals. It could be the signals of synchronization, Hello-packets, and so on (Fig. 7.4).

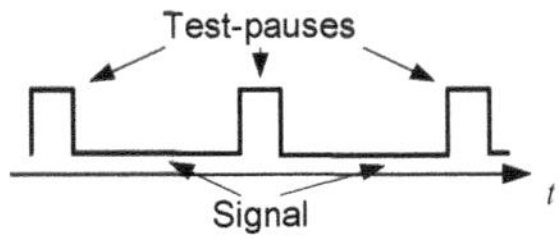

Fig. 7.4 Relation between useful signal and test-pauses

2. Two antennas are chosen. The main lobe for one of them is oriented on the useful signal. The main lobe of the second is oriented on the interference. Of course, it is possible the interference direction is known. The output of the second antenna creates the reference channel which is now free from the useful signal: $y_2 = y_{ref} = u_{ref} + v_{ref}$.

3. Two identical antennas are chosen (Fig. 7.5). Both antennas are oriented on the useful signal. The directions are chosen in such a way that amplitudes of these signals are equal, as well as their phases. Signals from both antennas enter the input of additive-differential unit (Σ/Δ). The additive output Σcorresponds to the primary channel; the differential output Δ corresponds to the reference channel. If the receiving channels are identical, then the useful signal $c(t)$ is mutually compensated on the differential output. The interference $n(t)$ differs from zero for both primary and reference outputs. It is true because the signal and interference cannot arrive from the same direction. Therefore, there is a difference $\varphi = \frac{2\pi d}{\lambda} \sin\theta$ between the interference phases in the first and second channels. In this formula the part $d \sin\theta$ is the difference of ray paths, whereas the part θ is the difference of their arriving angles (Fig. 7.6). In these figures, the antennas are denoted as $A_1$and A_2, whereas the distance among the antennas is equal to d.

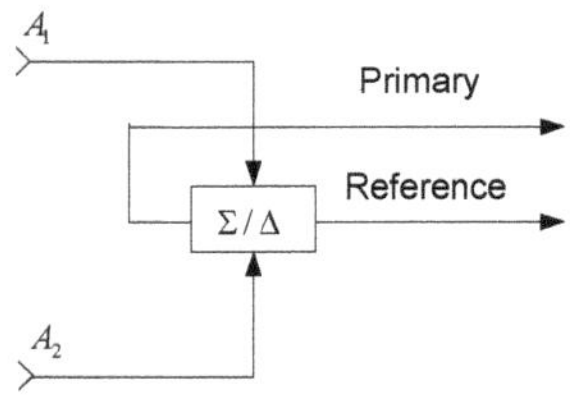

Fig. 7.5 Organization of outputs of antennas

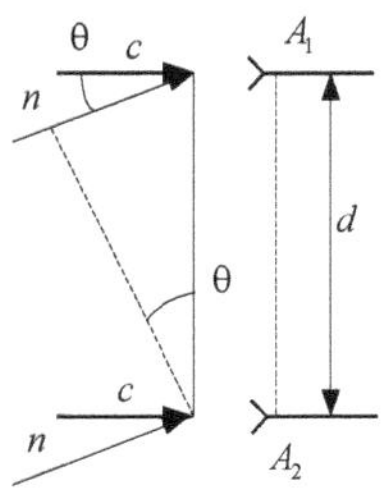

Fig. 7.6 Diagram of arriving for ray paths and interferences for antennas

Analysis of AIC operation can be done using mathematical models where it is easy to make changing for initial data. Next, this analysis should be justified by results of verification nature tests. The following conclusions can be made from this analysis.

1. It is useful to create the levels ratio for signal and interference as large as possible (in the reference channel). It is useful because the noise from this channel enters the primary channel. After multiplication by the weight coefficient W, this noise is summarized with the noise of the primary channel ν_{pr}. If in this case the following relation $n_{ref} > n_{pr}$ is true, then there is $W < 1$ and the level of noise ν_{ref} diminishes.

2. The useful signal can penetrate the reference channel. Due to correlation with the first summand of (7.26), the integration $\int AU_{ref}dt$ leads to compensation of the useful signal itself. Therefore, it is necessary to prevent penetrating reference channel by useful signal.

3. As a rule, the complex weighting coefficient $\hat{\dot{W}}$ is represented as four components and each of them is a quadrature. It means that four real components are estimated.

7.5 Examples of Practical Application of Control Algorithms in AIC

Let us discuss a two-element antenna with two interference compensators. As follows from above discussed issues, one of the most important problems for AIC is creation of qualitative reference channel. Let us consider one of constructive circuits of AIC (Fig. 7.7). It can be viewed as a development of the structure shown in Fig. 7.5.

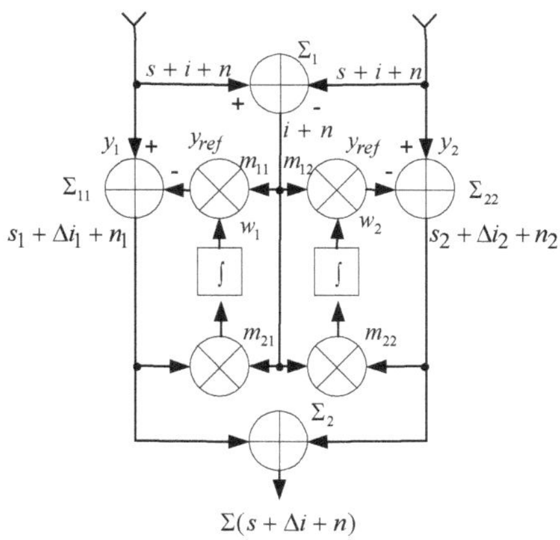

Fig. 7.7 Structure of AIC with double compensation of interferences

The essence of operation for this circuit is the following one. The antennas are in the area of equal phases of the useful signal as it is in Fig. 7.5. It allows to get the reference signal $y_{ref} = i + n$ as the output of the differential block Σ_1. Next, the signal y_{ref} enters two AIC placed in the receive channels of antennas A_1 and A_2. Next, the operation is the same as it was considered before. One of compensators includes the adder Σ_{11}, which makes compensation of weighted sum $y_1 - w_1 y_{ref}$ from the reference channel. It leads to the result of summation $\kappa_1 = s_1 + \Delta i_1 + n_1$. This very procedure takes place in the second AIC, where the result $\kappa_2 = s_2 + \Delta i_2 + n_2$ is formed. Obviously, the signals κ_1 and κ_2 are coherent in respect to the useful signal $c(t)$. It is true because they are coherent for inputs and outputs of the antennas. It allows their adding using the mutual adder Σ_2. The result of addition is the following one:

$$\kappa_1 + \kappa_2 = s_1 + s_2 + \Delta i_1 + \Delta i_2 + n_1 + n_2. \tag{7.28}$$

After addition, the level of coherent signal increases in four times. It corresponds to the equation: $(s_1 + s_2)^2 = s_1^2 + s_2^2 + 2s_1c_{s2} = 4s^2$. In the same time, the noise level increases only in two times. Really, the following equation can be formed $(n_1 + n_2)^2 = n_1^2 + n_2^2 + 2n_1n_2$, where $2n_1n_2 \to 0$ due to incoherence. Thus, we can state that the signal/noise ratio increases in two times in the given AIC in comparison with the trivial AIC. Of course, it is true if we neglect the interference remainder.

The second example is an adaptive watch system. If there is a possibility in use of some model (standard) signal y_m, then the control task is reduced to following this standard. In this case, the error (mismatch) signal $\kappa = \dot{W}_t y_t - y_m$ (it is a misalignment) is formed. Here the value of coefficient $\dot{W}_t$ reduces the misalignment to zero (in square-mean).

The algorithm for watch system is the same as for AIC, namely there is:

$$\hat{W}(t) = A(t)\hat{W}(t) + K(t)y(t)\left[W(t)y(t) - y_s\right]. \tag{7.29}$$

The block diagram of watch control algorithm for observations of a standard signal is shown in Fig. 7.8.

Let us discuss the mode of this system operation. It is necessary to have the output of watch algorithm $y(t)$ corresponding to the model (standard) signal. It means that it is necessary to watch a model signal. The output of adder Σ_1 generates the value of misalignment $\kappa = Wy - y_m$. Next, this value enters the input of multiplication unit m_2, which generates the following value

$$y(t)\cdot\kappa(t) = (s+v)(\Delta s + v) = \Delta ss + \Delta sv + sv + vv. \tag{7.30}$$

After integration, the value of $\int \Delta ss dt$ is used as a control influence for constructing weighting function $\dot{W}(t)$. Thus, the output $y^*(t)$ contains the signal coinciding with model signal y_m.

Let us discuss the adaptive antenna array (AAA). The idea of AAA is the following one. It is necessary to add weighted interferences received from n antennas using the mutual adder Σ_m. The summation should be executed in such a manner that the

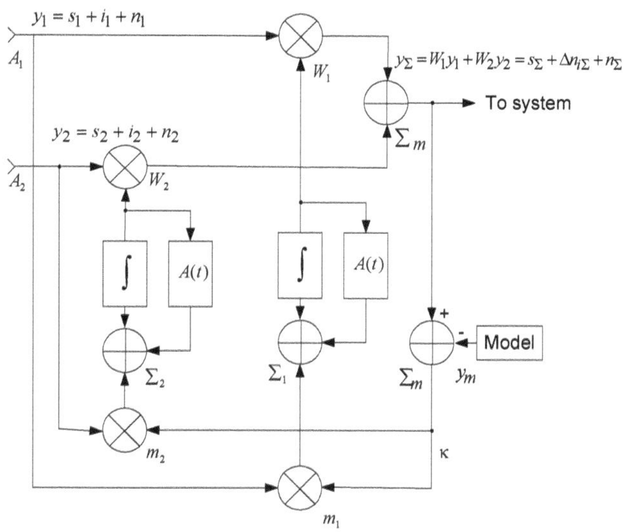

Fig. 7.8 Block diagram of watch control algorithm with standard signal

total interference influence is equal to zero. The block diagram of two-element antenna array is shown in Fig. 7.9. The adder Σ_m generates the misalignment κ, which contains $\Delta s_\Sigma + i_\Sigma + v_\Sigma = \kappa$ with $\Delta s_\Sigma \to 0$. It leads to subtraction of the useful signal and κ is used as the reference channel.

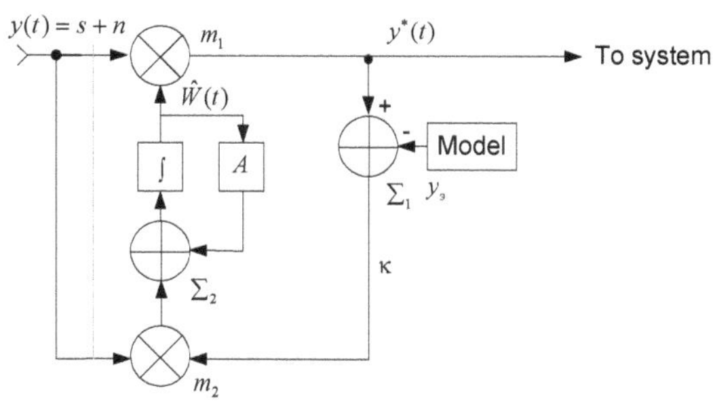

Fig. 7.9 Block diagram for algorithm of two-element AAA

7.6 Usage of Obtained Results in Modern TCS

The discussed algorithms of optimal control are introducing now into modern telecommunication system. They are used for control either functional or structural characteristics of TCS. The functional control can be spread for all five layers of control model (pyramid) represented in the TMN conception. The stable trend is observed to increasing the relative density of optimal control algorithms in the area of control of both structure and function of separate network elements. But there is no positive answer for the following question: is it possible to implement the global optimal automatic control method for a distributed telecommunication system? The future will show the answer, now it is not clear.

The policy of restructuring in TCS can be implemented using different approaches. If the direct approach is used, then the system adaptation is executed in the real time mode taking into account obtained estimates of functional characteristics. It leads to this or that alteration of the system structure: the by-pass paths are formed in the cases of failures, or available resources are redistributed if there are alterations in traffic, or necessary routes are created, or the route map is corrected, and so on. If indirect control method is used, then the adequate model of TCS is created. Its identification and adaptation is executed right along, using the obtained estimates. The model and structure of TCS are represented either as the state tensor, or by graph-analytical methods, or with incidence matrix, and so on. If the state of TCS is changed, that the restricting is executed using some optimization methods (for example, method of Ford–Fulkerson, or Bellman–Ford, or other).

The transit states of the structure can be taken on different calculation values. In the simplest (binary) variant, when these values can be equal either 0 or 1, the values form the usual incidence matrix. Increasing the number of variants gives possibility for discrete estimation of the loading for this or that communication direction. Due to standardization of rates for digital flows, it is possible to make choice according with existed standards. It allows making redistribution of loading (as well as other control actions) precisely enough.

The value of delay in the control loop is a very important parameter of TCS. As analysis shows, a control procedure reaches its goal if control cycles are significantly less than the period of state changing for a network requiring restructuring.

The considered control methodology for telecommunication systems is general enough. It allows efficient solutions for problems of changing the network structures. It can be applied for all types of networks: hybrid, convergent and multiprotocol. It is true due to the fact that adaptive control procedures are practically free from peculiarities of this or that technology.

Recommended Literature

1. Dorf, R., Bishop, R.: Modern Control Systems. Prentice Hall PTR, Englewood Cliffs (2003)
2. Kwakernaak, H., Sivan, R.: Linear optimal control systems. John Wiley and Sons, Chichester (1972)
3. Sage, A., Melsa, J.: System Identification. Academic Press, London (1971)
4. Singh, M., Titli, A.: Systems: Decomposition, Optimization and Control. Pergamon Press, Oxford (1978)

Chapter 8
Taught Controlled Systems

Abstract. The chapter discusses the topic of taught controlled systems. Two kinds of taught systems are considered. First of them is a learning by instruction, whereas the second has no teacher (instruction). There are some examples of practical implementation of these approaches in the existed telecommunication systems. The following classes of taught systems are analysed: the systems with identification of a model, the systems of the search type, and the self-organized and self-repairing systems with re-engineering. The entropic, homeostatic and morphogenetic solutions are discussed.

8.1 General Taught Principles for Dynamic Systems

When the cybernetics appeared as a science, the possibility appeared not only for formalizing different systems, but for equipping them some anthropomorphous (or human) property. The term "artificial intelligence" was coined; adaptive algorithms appeared, as well as decision makings under risk, self-organization, taught systems, and so on. These terms connected with the same systems reflect this or that property of controlled system, operating under conditions of uncertainty. We think the most general term here is "taught controlled systems", because all above mentioned systems include some taught procedure.

The taught system is a system operating under conditions of uncertainty and accumulating information about parameters of this uncertainty to use this accumulated information for solution of general-system control problems. Two main methods can be distinguished between different taught approaches. First of them is learning by instruction (with a classified taught sequence). The following issues can be use as an instructor: model signals, test sequences (test pauses, test signals), and hello-packages. The hello-package does not carry any information components; they are used for finding parameters of delays, transferring verification, and so on. The second approach is the learning without instruction; it is executed without the taught sequence.

V. Popovskij et al.: Control and Adaptation in Telecommunication Systems, LNEE 94, pp. 97–112, 2011.
springerlink.com

Let us discuss the methods of learning by instruction. If an instructor is presented as a model signal y_m, then the learning can be organized with using the following statistical procedures:

1. Accumulation of statistics about the general amount of received signals and about amount of correctly received signals n_c. An unknown probability P_{err} of incorrect receiving can be found, where there is:

$$P_{err} = \lim_{n \to \infty} \left(\frac{n_\Sigma - n_c}{n_\Sigma} \right). \tag{8.1}$$

2. By analogy with the probability (8.1), it can be found the probability of lost packets, the probability of readdressed packets and packets with errors, and so on.

3. Accumulation of statistics about unknown parameters with following processing this statistics to obtain sample mean values of these parameters, variance, correlation intervals, distributions of probabilities and other characteristics. Next these characteristics are used in optimization procedures (the identification task).

Therefore, the statistics methods are used for obtaining this or that unknown characteristic. It is treated as the learning procedure.

Existence of a model signal can be directly used in different recursive procedures in real time scale, when there is no statistics accumulation. This approach can be used in: 1. The standard procedure of Kalman–Busy filter for generation misalignment $(y(t) - y_m)$.

2. The MIMO (multi-input, multi-output) technology for finding unknown transfer coefficients h_{ij}. In MIMO technology the test signals are transferred one by one between an antenna A_1 and an antenna A_2. These signals are used for obtaining the estimates $\hat{h}_{ij}$. Next these estimates are substituted into watch equations y_1 and y_2. This approach is illustrated by Fig.8.1.

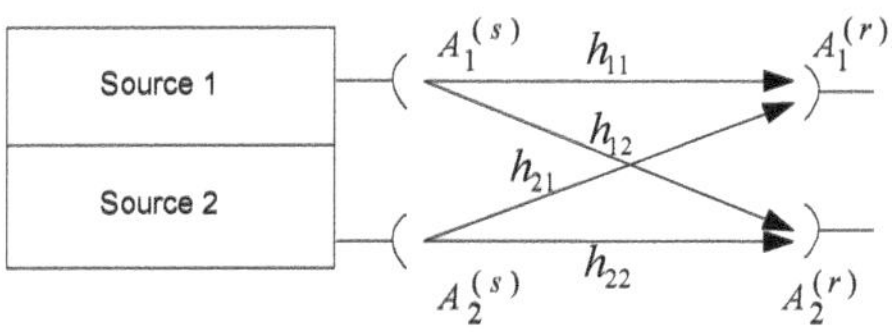

$$\left.\begin{aligned} y_1 &= \hat{h}_{11}x_1 + \hat{h}_{21}x_2 \\ y_2 &= \hat{h}_{12}x_1 + \hat{h}_{22}x_2 \end{aligned}\right\} \tag{8.2}$$

Fig. 8.1 Transmission in MIMO technology

If the values of $\hat{h}_{11}, \hat{h}_{12}, \hat{h}_{21}, \hat{h}_{22}$ are known, then it is easy to solve the system of linear equations (8.2) relatively variables x_1 and x_2, where x_1 and x_2 represent two independent data flows. Two source antennas ($A_1^{(s)}$ and $A_2^{(s)}$) are used in MIMO technology for transmitting, whereas two antennas ($A_1^{(r)}$ and $A_2^{(r)}$) are used for receiving. Also, there are four receiving directions with four different transfer coefficients $h_{11}, h_{12}, h_{21}, h_{22}$. The Multiple input-Multiple output technology allows transmitting two independent digital flows having the same frequency. In this case, there is the multipath propagation of radio waves, but it does not prevent the qualitative receiving. Moreover, it plays a positive role because coefficients $h_{ij}(t)$ stay independent and they vary permanently. To get the required estimate $\hat{h}_{ij}$ (for learning) the test-signals are used. These signals are transmitted through the time $\Delta t << \tau_{cor}$, where τ_{cor} is the correlation interval of a multipath faded signal. In reality, there is $\Delta t \approx 0.01\text{sec} - 0.1\text{sec}$. The main advantage of the learning by instruction is higher reliability of obtained results, than in the alternative method. It is true if there is a classified learning sequence. The disadvantage of this method is the decrease of throughput due to necessity in transmitting test-signals.

Let us discuss the methods of learning without instruction. This kind of learning is implemented using synthetic (implicit) data. The information sequence by itself can be used for learning. This method is less effective because tremendous errors are possible in learning. For example, an error in transmission can lead to the error in learning. Therefore, there is no sense in the learning without instruction if there is a low reliability of a channel in use. On the other hand, this approach simplifies a learning algorithm and does not require losses of time for transmitting test-signals.

The block diagram of the adaptive self-taught system $S(x,u,t)$ can be represented as composition of two methods, combining approaches of Ponselle and Watt. The block diagram is shown in Fig. 8.2.

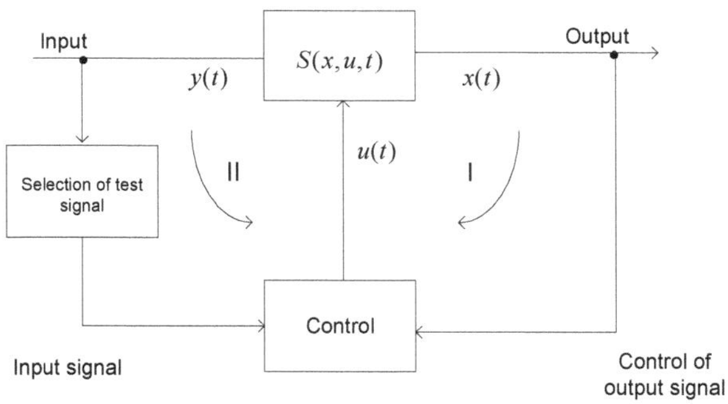

Fig. 8.2 Block diagram of adaptive self-taught system

In contrast with above mentioned single-loop control circuits, the adaptive system includes two loops. First of them executes the control by deviation (back control). The second implements the control by influence (front control). In this approach (Fig. 8.2) the system control is generated on the base of the value of mutual correlation of input and output signals.

8.2 Classification of Taught Systems

The taught system can be divided by the following classes: taught system with model identification; taught systems of search type; taught self-organized systems, and self-repaired systems with reengineering. Let us discuss some peculiarities of these systems.

Let us start from the taught systems with model identification. Actual parameters of systems do not always coincide with a model used in an algorithm of estimation and control. For example, in the following equation of the model state $dx(t)/dt = A(t)x(t) + B(t)u(t) + C(t)\xi(t)$ the parameters A, B, C can differ from the actual situation. It means that, for example, the actual value of A can be found as $A_{act} = A_{opt} + \Delta A$, where the symbol ΔA stands for deviation of the model due to its inertial properties. The correlation interval τ_{cor} of the model in use can be represented as $\tau_{cor} = \tau_{opt} \pm \Delta\tau$. If a model with higher performance is chosen, where there is $\tau_{cor} > \tau_{opt}$, then the control system will react on the small influences, such as a noise. If a model with lower performance is chosen, where there is $\tau_{cor} < \tau_{opt}$, that not all influences will be followed up and the smoothing effect will arise.

If there is a deviation for coefficient B, where there is $B = B_{opt} \pm \Delta B$, then the control can be either with overreaction (I) or with undershoot (II) (Fig. 8.3).

The value of $C(t)$ shows the level of generated process $x(t)$. If the value of $C(t)$ is too high, then the unstable mode of KBF can occur in practice.

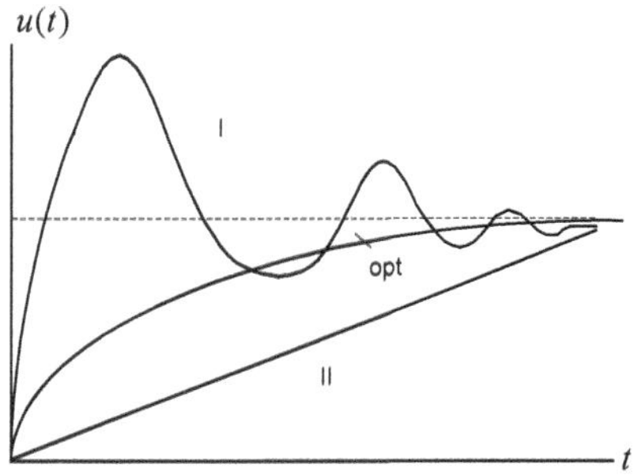

Fig. 8.3 Transient modes of controlled system under nonoptimal choice of model parameters

Therefore, it is necessary to get estimates of unknown parameters, if there is uncertainty with system parameters. In other words, it is necessary to identify the actual model. To find the estimates of unknown parameters, it is necessary to construct some additional procedure, which is independent from the main procedure. This additional procedure can be either RM filter or KBF. The main and additional procedures can operate in parallel. In this case, a vector of estimated parameters $\vec{x}^T = (x, A, B, C)$ can be formed. Next, the estimation task for the vector $\vec{x}$ should be solved using the additional procedures. The obtained estimates $\hat{A}, \hat{B}, \hat{C}$ are substituted instead of unknown parameters into units executing the main estimation of $\hat{x}$ and control (Fig. 8.4).

For example, the identification algorithm corresponding to the Robbins-Monro procedure is represented by the following equation:

$$\hat{A}(k+1) = \hat{A}(k) + K(k)\left[y_A - A(k)\right]. \tag{8.3}$$

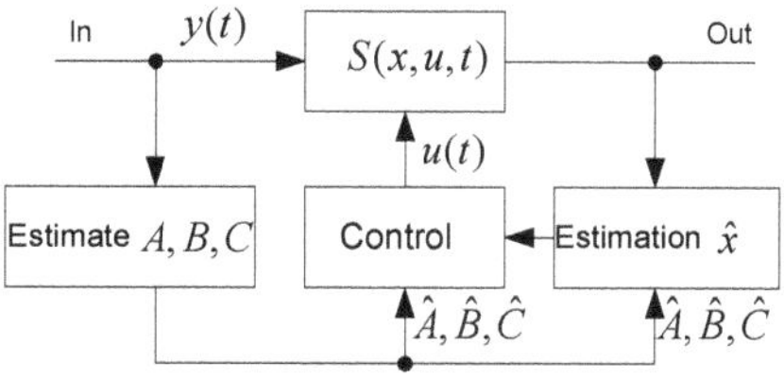

Fig. 8.4 Block diagram of controlled system with identification of coefficients $\hat{A},\hat{B},\hat{C}$

In (8.3) the equation $y_A = A(k) + \nu(k)$ reflects the result of measurement of the parameter $A(k)$. So, it is a watch equation for a parameter to be identified. The block diagram and sequence of actions needed for identification the value of parameter $A(k)$ is shown in Fig. 8.5. As you can see, it includes three main blocks. One of them is used for measurement of $A(k)$. Its output is connected with the block of estimation of the required parameter. Next, both the function y_A and result of estimation are processed by the main block.

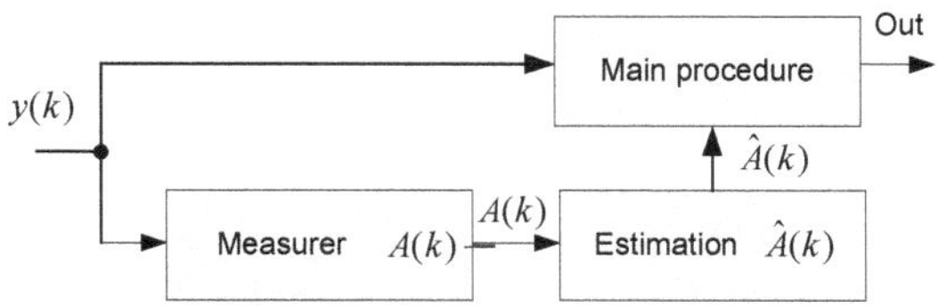

Fig. 8.5 Sequence of execution of procedures under identification of parameter $A(k)$

Let us discuss more thoroughly the algorithm of estimation for the vector $\mathbf{x}^T = (x,A,B,C)$ under its discrete representation. The state equation for the correlation interval τ_{cor} is represented as the following one:

$$\tau_{cor}(k) = A^{-1} = x_\tau(k) = \Phi^{(\tau)}(k,k-1)x_\tau(k-1) + G^{(\tau)}\xi_\tau^{(k)}. \tag{8.4}$$

In (8.4) the function $\Phi^{(\tau)}(k,k-1) = e^{-\frac{\Delta t}{\tau_k}}$ is a predictive function for alterations of the correlation interval τ_{cor} with the correlation interval of these changing τ_k. The coefficient $G^{(\tau)} = \sqrt{\hat{N}_\xi(1-e^{-\frac{\Delta t}{\tau_k}})}$ is a intensiveness coefficient for the generated random process $\xi_\tau^{(k)}$, whereas the estimate $\hat{N}_\xi$ is an estimate of degree of spectral density for the process rate $\xi(k)$.

As a result of substitutions into the state equation (8.4), the correlation interval τ_{cor} is represented as the following one:

$$\tau_{cor}(k) = e^{-\frac{\Delta t}{\tau_k}}\tau_{cor}(k-1) + \sqrt{\hat{N}_\xi(1-e^{-\frac{\Delta t}{\tau_k}})}. \tag{8.5}$$

The equations for either spectral density N_ξ or coefficient C are made by analogy with the equations (8.4) and (8.5). As a final result, the following system of equations can be obtained for the estimation algorithm for the vector $\overrightarrow{x}$:

$$\hat{x}(k) = \begin{cases} x(k) = e^{-\frac{\Delta t}{\tau_{koo}}} x(k-1) + \sqrt{\hat{N}_\xi (1 - e^{-\frac{\Delta t}{\tau_{koo}}})} \xi(k), \\ \tau_{kop}(k) = e^{-\frac{\Delta t}{\tau_k}} \tau_{kop}(k-1) + \sqrt{\hat{N}_\xi (1 - e^{-\frac{\Delta t}{\tau_k}})} \xi_\tau(k), \\ N_\xi(k) = e^{-\frac{\Delta t}{\tau_N}} N_\xi(k-1) + \sqrt{\hat{N}_\xi (1 - e^{-\frac{\Delta t}{\tau_k}})} \xi_N(k). \end{cases} \quad (8.6)$$

In the system (8.6) the first line corresponds to the main equation. It uses the estimates $\hat{\tau}_{cor}$ and $\hat{N}_\xi$. The second line contains the state equation for $\hat{\tau}_{cor}$, whereas the third line is a state equation for $\hat{N}_\xi$. As follows from the equations of the system (8.6), the algorithms for estimation of the vector $\overrightarrow{x}(k)$ are made using the standard Kalman–Busy procedures (6.16).

This adaptive procedure for model identification is complex enough, especially if a vector $\overrightarrow{x}$ has many dimensions. There are simpler (but maybe less effective) adaptive methods. One of them is a method of locally stationary approximation. Let us discuss this approach.

The method of locally stationary approximation is based on the assumption that any nonstationary telecommunication process can be represented as a stationary one for some time intervals ΔT (Fig. 8.6). It is presumed that there are an algorithm and a device capable to find the boundaries of these intervals ΔT. After finding these local parts, the main algorithm of estimation and control executes the correction of corresponding nonstationary parameter (or parameters).

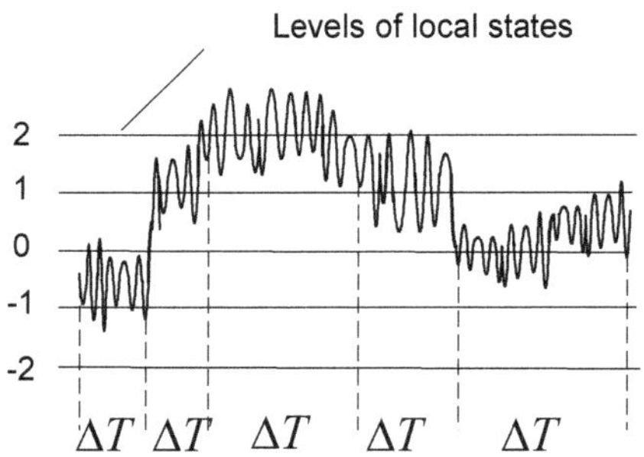

Fig. 8.6 Partition of nonstationary process by parts of local stationarity

Very often, one from parameters turns to be nonstationary. For example, it can be the arrival rate. In this case the algorithm of estimation for losses in quality of functioning primary device can be used as an indicator of transition time from one local-stationary part to other. The block diagram of adaptive algorithm with using parts of local stationarity is shown in Fig. 8.7.

As follows from the above discussed approaches, it is necessary to collect this or that statistics about states of different modes of network elements to be able to solve these or those adaptive problems. In telecommunication system, the SNMP protocol can be used for collecting necessary data about states of network parameters. This protocol provides control tools and it controls of network elements, configurations, performance and safety. It provides gathering statistical data about readings for different parts under availability of agents providing transmission of necessary

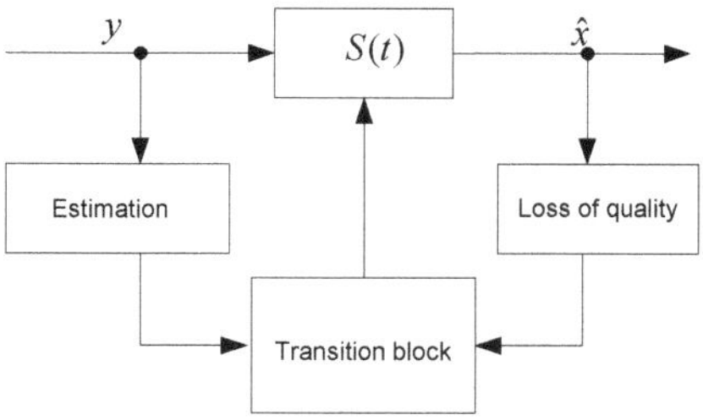

Fig. 8.7 Block diagram of adaptive algorithm with using parts of local stationarity

information. The accuracy of identification procedure depends on many parameters. One of the most significant is the delay time for the control channel. Let us discuss the influence of this factor.

As it is known, the telecommunication system is a technical-organizational system distributed in both time and space. It can have different scales, such as LAN, MAN, WAN. Processes in TCS take places in geographically distributed segments (parts). These processes are interrelated. It means that TCS is characterized by different values of delay time Δt_d. But even insignificant values of delay times Δt_d result in delayed reaction of the taught subsystem. This delay results in loss of learning quality. Therefore, the quantity of delay should be comparable with the value of correlation interval. The explanation to finding the correlation interval is shown in Fig. 8.8.

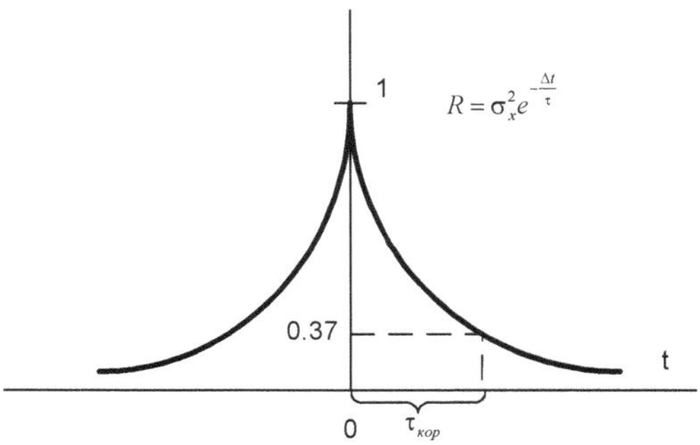

Fig. 8.8 Explanation to finding the correlation interval

The delay is a metric used for control of priorities and routes. The delays under interrelations (such as information interchange, control, and monitoring) are stipulated by the four following factors. Firstly, there is a delay in the information interchanges from one element to another due to finiteness of signal propagation in communication channels. Secondly, there is a delay which is caused by queues and buffer devices (a queue delay). Thirdly, there is a delay connected with processing and transformation of information (encoding, decoding, estimation, and so on). Fourthly, there is a delay connected with latency in the time of obtaining permission for transmission.

The existence of delays leads to loss of quality of transmission. But it also shows under the solution of the following four control tasks. Firstly, it diminishes the quality of the task of the remote monitoring. Secondly, it has a negative influence on the time of gathering information about the states of system channels. This information is called LSA (link state advertisement); it is used in routers for renovation of the route tables. Next, it influences the flow control using the method of sliding window flow control. The essence of this method is in permission of a receiver to transmit some data for filling a window. At last, the delay influences the flow control because of the states of buffers in receivers. For example, the procedure RED can ban the transmission till the buffer is busy, and so on.

The delays Δt_d lead to errors in estimation of unknown parameters and, therefore, in implementation of control procedures. The delay Δt_d is a part of the prediction function $\Phi(k,k+1) = e^{-\frac{\Delta t+\Delta t_d}{\tau_{cor}}}$. The function $\Phi(k,k+1)$, in turn, is a part of the equation for estimation of the discrete state:

$$\hat{x}(k+i) = \Phi(k,k+1)\hat{x}(k) + K(k)\left[H\hat{x}(k) - y(k)\right]. \tag{8.7}$$

Therefore, existence of the delay Δt_d is equal by its influence to increase the digitalization step from Δt till $\Delta t + \Delta t_d$.

8.3 Adaptive Search Taught Systems

The search systems can be divided by two main classes. The first of them includes the object search systems aiming in the finding some specific object (a model). The second class includes the extremal search systems aiming in the finding either minimum or maximum values of something.

The tasks of the object search appear under the loss of a subscriber either in some frequency range or in the space, or under the loss of synchronization and in the search of synchronization signal, and so on. To solve the search problem it can be used either the method of direct enumeration of possibilities (the direct search) or the dedicated search (goal seeking) using some attribute. As a rule, some search mechanism is included into a search system. This mechanism includes such components as the search algorithm, the signal attribute used in the search, the algorithm for detection of the lost signal, and a control unit providing keeping from the next lost of the detected signal.

Search algorithms are constructed taking into account both structure and the number of dimensions of the search space. One of approaches is the linear search. It is used, for example, to find a lost signal in some frequency range.

There is the following sequence of operations under the linear search. Firstly, the frequency range is divided by intervals Δf (Fig. 8.9a). The intervals Δf should be such that it is possible to detect a signal into the frequency band $\Delta f_i \approx \Delta f_c$. Next, the search procedure is executed for each interval till a desired signal will be found. The values Δf_i should be chosen from the condition $\Delta f_i \approx \Delta f_c$. If this condition is satisfied, then it minimizes the possibility for the lost of the signal. Let us point out that the search time for each interval should be chosen from the condition that all transient processes should be finished and the necessary processing should be executed.

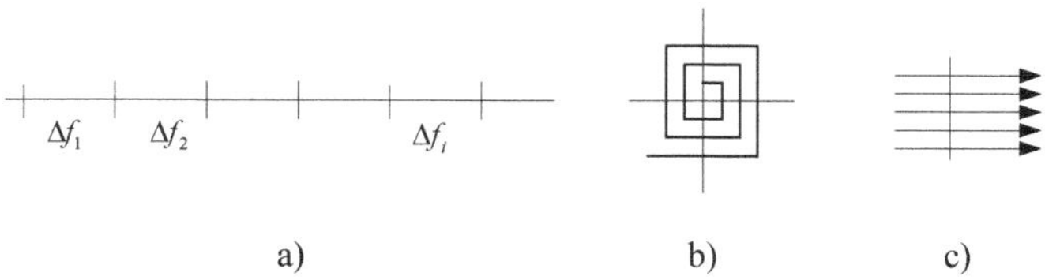

Fig. 8.9 Explanations to search algorithms

The search inside some area, as a rule, is executed with help of the spiral motion (Fig. 8.9b). The search in parallel channels can be executed either for total area (Fig. 8.9c) or sequentially (as it is done for scanning in the TV sets).

It is very important to find a correct signal attribute to organize the signal search. A detectable signal should be different from all other signals. It could be either difference in the signal structure or in some other characteristics. For example, the detectable signal can be a broadband signal (for example, the Barker signal), or it can be represented by some *B*-sequence, and so on. The chosen signal attribute should increase the difference between the detectable signal and other signal existed in the search space. The detection algorithm can be built on the base of selection a hypothesis from some alternative. For example, it can be similar to algorithms used for receiving of conventional information signals.

A control unit can be organized using the principle of model monitoring. In this case the misalignment $[y - y_m]$ is organized and, next, either the RM procedure or KBF is used for estimation. As an example, let us discuss the problem of control by the directivity diagram (DD) of some mobile counterpartner. It is the usual task of target following (Fig. 8.10). The solution can be obtained either with help of the electric actuator of the antenna (due to electromotor) or using some electronic methods. One of the electronic methods is usage of the adaptive antenna array and amplitude-phase distribution of its elements. This solution can be oriented either on the search of equisignal zone for the differential directivity diagram (Fig. 7.10a) or on the search of the maximum of receiving (Fig. 8.10b). In the first case the target direction is found due to rocking of DD in the sector $\Delta\varphi_{1,2}$.

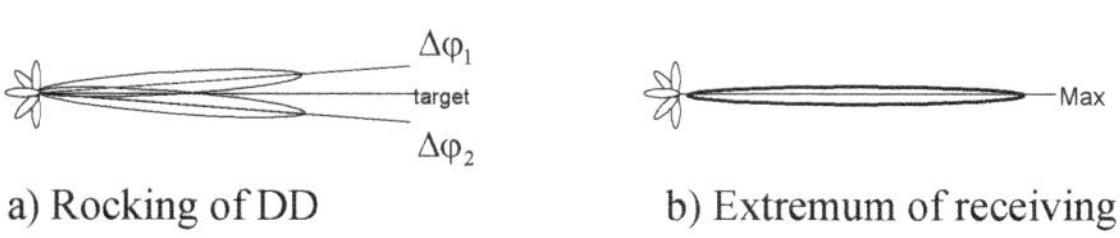

Fig. 8.10 Control of antennas for target following

8.4 Taught Adaptive Self-organized Algorithms in Controlled Systems

Three different types of self-organization are distinguished in the general system theory. The processes of self-generation of organization form the first class. The second class includes the processes which help supporting some level of organization under the alterations in internal and external conditions of functioning. At last, there is a third class including the processes connected with perfection and self-evolution of the systems capable for accumulation and usage of their past experience. With regard to telecommunication systems, the second class of processes is characteristic. But it is quite possible that two other processes will find their place in the nearest future.

The self-organization is a category dealing with functional characteristics of a system. At the same time, structural methods are widely used for the goals of self-organization. As a result of operation of some dedicated (task-oriented) system, these or those changing can take places under influence of internal and/or external conditions. These changing can reach some defined critical level and they are stimuli for correction of methods used for control of functional and structural characteristic without changing the purpose of a controlled system.

The discussed above controlled adaptive algorithms are mostly used for the correction of system characteristics. The adaptation is possible either for some system elements or for the total system. The characteristic example can be the adaptation of a network to the traffic changing. But operating conditions can be changed in such a manner that it is impossible to continue operation using only available corrective tools. In this case the system can use a mechanism of the purpose changing (or changing of a criterion).

Let us discuss the content of the tasks for adaptive and self-organized systems. A dedicated controlled system $S(x,u,t)$ is named adaptive if it includes some procedure (algorithm) providing efficient operating under conditions of uncertainty.

Obviously, this definition is not unambiguous, as well as the conception of adaptability. It requires some further discussion. In the given above definition, it is supposed that there is some controlled system $S(x,u,t)$, which is optimal in the boundary of some specific restrictions (constrains) Y. It is clear, if there is either change of constrains or appearance of new additional conditions Δy, then the system cannot be treated as optimal under the new conditions $Y' = (Y, \Delta y)$. But if the system $S(x,u,t)$=$S(t)$ includes some procedure $a(t)$, which provide optimizing the system $S(t)$, then now we deal with a new, modernized and optimized system $S'(t) = S(a,t)$. This system can be treated as adaptive in relation to both the changing Δy and the system $S(t)$.

Continuing our reasoning, it is possible to state that the system $S'(t)$ is a dedicated controlled system optimal under conditions Y'. But these conditions can be changed further till there are some new conditions Y''. In those new conditions the system

$S'(t)$ is not optimal and it needs introducing some new procedure $a''(t)$ providing optimization for the conditions Y''.

It follows form our discussion that the conception of "adaptability" is as relative as the conception of "optimality". Let us introduce one more definition leading to better understanding of the essence of the adaptation. The adaptation is a process of infinite optimization of a controlled system under conditions of variable and random in time external actions (influences). Let us discuss the methods used for constructing adaptive procedures.

There are two main methods of adaptation used in controlled systems. The first of them is the adaptation to conditions of the system $S(a,x,u,t)$, when it is necessary to correct either parameters of some system elements or the structure. This necessity is connected with some changing in external or internal conditions of operation.

The second approach is the adaptation to the condition in the observation channel. It can be represented by the following equation:

$$y(t) = Hx(t) + Dn(t) + v(t). \tag{8.8}$$

The essence of the adaptation procedure in the observation channel is reduced to the following. If there is an interference $Dn(t)$ in the observation channel, then it is reasonable to choose an observation basis minimizing the influence of this interference.

The first approach is implemented using identification of internal and/or external conditions of the system operating. In Section 8.2, it was mentioned that the identification methods target suppressing of a priori uncertainty in controlled systems. Because of it, the systems with identifications are treated as adfaptive and it coincides with the definition given above. The peculiarities of the second approach are discussed in Chapter 7.

8.5 Constructing Procedures of Self-organizing and Self-repairing

The self-organized and self-repairing systems can be treated as a further development of adaptive systems. Let us explain the essence of these systems.

An adaptive system $S_a(t)$ can decrease the quality of its operating under influence of some internal or external factors. In this case, some tools should be provided for correction of functional and/or structural system properties. These tools target improving the operating quality using some purposeful enumeration of corrective possibilities. As a rule, the change of the system objective (criterion) is not assumed during the stage of enumeration of variants. Obviously, the move from the adaptation mode to the mode of enumeration of corrective possibilities means the transition to the new type of the systems, namely to self-organizing systems.

The mode of self-organization is connected with the dedicated search process of new operating modes for either some elements, or their groups, or the whole system. Therefore, some new variants of functional properties are looked for, as well as the choice of variants for restructuring. The restructuring is reduced to finding more rational structures. In this case such phenomena are possible as appearance or disappearance connections between the system elements, as well as appearance of new elements or new properties of existed elements. The solution is also possible which is viewed as some alternative to previous approaches. If there are large amounts of available resources and requests for their use, then they are included in the solution of general system tasks for providing the desired service quality. Such a solution is named an entropic approach. In practice, it is used in its pure form only under critical situations. By analogy with the principles of stability, the homeostatic methods can be used as an addition to the entropic approach. The homeostatic method is reduced to the choice of correction variant for both functional characteristics and operating modes of the network elements. The second important method is morphogenetic, based on the system restructuring.

The entropic method is used relatively seldom. But the second two approaches (homeostatic and morphogenetic) are used, as a rule, in the case of the first necessity, when a controlled adaptive system does not reach required level of operation.

Let us discuss the peculiarities of the enumeration of variants under solution of the problems connected with the self-organizing of adaptive systems. It is necessary to foresee a set of variants and programs of their execution, as well as a strategy for enumeration of these variants. In the same time it is necessary to develop algorithms providing system moving from one variant to another. All these tools are necessary for solution of the tasks connected with the self-organization.

If functional characteristics are chosen, then the operating modes are changed for different network elements. It could be the rules of access, methods for distribution of the network resources, mechanisms of preventing congestions, and so on. If structural characteristics are chosen, then there are changes for methods of routing, rules of search and activation of abonent stations, access points and other network elements, and so on.

The first problem arising under the choice of variants is the following one: which characteristics should be taken for the beginning of the primary choice of variants. Obviously, the choice can start either from structural or from functional characteristics. The logic of choice is not unambiguous; it depends significantly on the priorities of manufacturing planners. In the most general case it is possible to try to make a parallel choice in the set of functional and structural variants. Let us discuss the possibility of formalizing the process of self-organization in details.

Let us start from the directions in formalization of the self-organization process. If the adaptation is a process targeting either some specific object or influence, then the self-organization process is directed on some integrative properties of a system. As well as both the control and adaptation, the self-organization process is a purposeful process needed corresponding optimization strategies under the restriction of available resources. The QoS (quality of service) characteristic can be chosen as an integrative characteristic determining the operation quality of telecommunication system. If such a choice is made, then the self-organization process can be viewed as some complex of actions targeted solution of the tasks of monitoring, control and administration of QoS. The self-organization based on QoS (let us name it as self-organization by QoS) includes the following actions: control of monitored parameters of QoS; alarm (warning) system; supporting the required level of QoS; demands about some information or actions of QoS; warning on the base of events related with control of QoS.

Therefore, the self-organization by QoS should comprise the decision-making tasks respectively to some factors combining of which form the qualitative characteristic of QoS. Besides, the self-organization is not a one-time procedure, it is rather a multi-phase procedure executed in the real time. Different criteria can be used for the decision-making process. Let us discuss the main criteria used on the self-organizing systems.

Let us start from the average gain criterion. The efficiency of a particular decision is evaluated as the mean expected value (mathematical expectation) of estimates K_{ij} for all states of an existed environment, namely:

$$K(x) = \sum_{j=1}^{m} P_j K_{ij}, i = 1,2,\ldots,m. \tag{8.9}$$

The optimal decision made about the self-organization of a particular system is determined by the following expression:

$$K_{opt} = \max_i \sum_{j=1}^{m} P_j K_{ij}. \tag{8.10}$$

The next criterion to be discussed is the criterion of careful observer (or Wald's criterion). It is a minimax criterion, which guarantees some gain under the worst conditions. The criterion is based on the following rule: if the system states are not known, then it is necessary to act carefully. It means that the orientation should be made on the minimum value of efficiency determined as

$$K_{opt} = \max_i(\min_j K_{ij}), i = 1,2,\ldots,m; j = 1,2,\ldots,n. \tag{8.11}$$

The minimax criterion presumes making of decisions which do not contain any risk elements. Therefore, in reality any decision can give a gain which is greater than the gain obtained for (8.11).

Next criterion in use is a maximax criterion. According to this criterion, the quality is estimated using the maximum value of efficiency. The optimal decision is the one having the best values for all K_{ij}. It can be expressed by the following equation:

$$K_{opt} = \max_i(\max_j K_{ij}), i, j = 1,2,\ldots \tag{8.12}$$

The criterion (8.12) is considered to be globally optimal.

Next popular criterion is a criterion of generalized maximin criterion (or Hurwitz's criterion). According with this criterion two values of efficiency are taken into account. The first of them is the highest ($\max K_{ij}$) and the second is the lowest ($\min K_{ij}$) values. The risk coefficient $\alpha (0 \leq \alpha \leq 1)$ is introduced to make their weighting. It determines the relation of this criterion with a decision making by a PMD. The efficiency is determined as the following weighted sum:

$$K(x) = \alpha \max_{i} K_{ij} + (1-\alpha) \min_{j} K_{ij}.$$

The optimality is determined by the choice of the maximum for all existed variants:

$$K_{opt} = \max_{i} [\alpha \max_{j} K_{ij} + (1-\alpha) \min_{j} K_{ij}] 0 \leq \alpha \leq 1.$$

If there is $\alpha = 0$, this criterion is reduced to the maximin criterion. If there is $\alpha = 1$ $\alpha = 0$, this criterion is reduced to the maximax criterion. Let us point out that the choice of the value for parameter α carries a subjective nature (it depends on the system developer).

The criterion of the minimum risk (or Savage's criterion) minimizes losses of efficiency under the worst conditions. In comparison with the criterion of Wald this criterion pays a bit more attention to the gain than to the loss. In can be represented by the following equation: $K_{opt} = \min_{i} (\max_{j} \Delta K_{ij})$, $(i, j = 1, 2, \ldots)$. In this equation the symbol $\Delta K_{ij} = \max K_{ij} - K_{ij}$ determines the difference between the maximum and current values of the state K_{ij}.

Therefore, the correct choice of a criterion gives the possibility for correct estimation of efficiency for some system in some instant of time. It gives the possibility of necessary correction of either structural or functional characteristic of the system. If a given system should always operate in the optimal way, it is necessary to correct its characteristics during the total time of this system's operating. It is the essence of the process of self-organizing. But it is impossible to change significantly the system characteristics for each correction step. Obviously, only some part of the system characteristics should be changed during one step of correction. Because of it, the procedure of self-organization is represented as some additive term ΔS to the main state of the system. It can be represented in the following way:

$$S(t_k, x_k, \sigma_k, K_{ij}^{(k)}) = S(t_{k-1}, x_{k-1}, \sigma_{k-1}, K_{ij}^{(k-1)}) + \Delta S(t_{k/k-1}, x_{k/k-1}, \sigma_{k/k-1}).$$

In this equation the symbols x_k, σ_k stand for the state and structure of the system for the step K correspondingly. The symbol $\Delta S(\cdot)$ stands for a part of the system which should be corrected.

It is possible to represent a self-organizing system by the following multiplicative equation: $S(t_k, x_k, \sigma_k, K_{ij}^{(k)}) = P_s(t_{k/k-1}) S(t_{k-1}, x_{k-1}, \sigma_{k-1}, K_{ij}^{(k-1)})$. In this equation the symbol $P_s(t_{k/k-1})$ represents the transitional probability, presuming the Markov's properties of the process of self-organization. Obviously, in this case the transitional probabilities should be determined for each state of the system, where the process of self-organization takes place.

8.6 Self-repairing of Controlled Systems

It is quite possible that applying procedures of adaptation and self-organization does not give back the desired quality of the system operating. In this case it is necessary to change the object function of the system to be corrected. This procedure can be named the method of self-repairing. Different style of actions is possible. The self-repairing can be done using either situational methods based on the experience of PMD or formalized methods using some program. Let us point out that the solution of the repairing problem for a degrade system belongs to the PMD. This PMD is a manager responsible for the quality of operation.

The problem of formalized approach for self-repairing of a controlled system is very complex and ambiguous. It is reasonable to refer to the Ashby's theorem about the solution of problems. It states that a solution of the problem having diversity should have even more diversity in itself. Because the telecommunication systems are now the most complex artificial systems, the ways for solving their problems are not always obvious (and trivial). But it is possible to list some typical solutions taken in the practice.

The simplest solution is merely the reducing of the level of required quality. This method is often used in practice when it is obvious that there is no possibility for keeping the required quality with available facilities of the system. The second approach is consisted into reformatting of the system. It presumes changing of some elements or interconnections leading to rise of the quality in comparison with the previous format of the system. These changing should affect the system-wide quality. The third approach is the changing of strategies. It concerns individual services of a system and coefficients determining the cost of these or those solutions affected the operating quality. The last approach is connected with changing profile of a controlled system. It means that some criteria, object functions and perspectives of the system should be chosen. Very often they differ from the corresponding issues of the system to be repaired.

A system $S(t)$ including three embedded systems (adaptive, self-organizing, and self-repairing) is shown in Fig. 8.11.

The formalization of self-organized and self-repaired systems is a very complex problem. It requires many researches connected with the order of execution of individual procedures, as well as with constructing of the system by itself. Obviously, the solution of the problems connected with self-organization and self-repairing is still a responsibility of a manager making decisions. But it is worth pointing out that some separate solutions find their application, for example, under the constructing of personal networks with the technology IEEE 802.15.

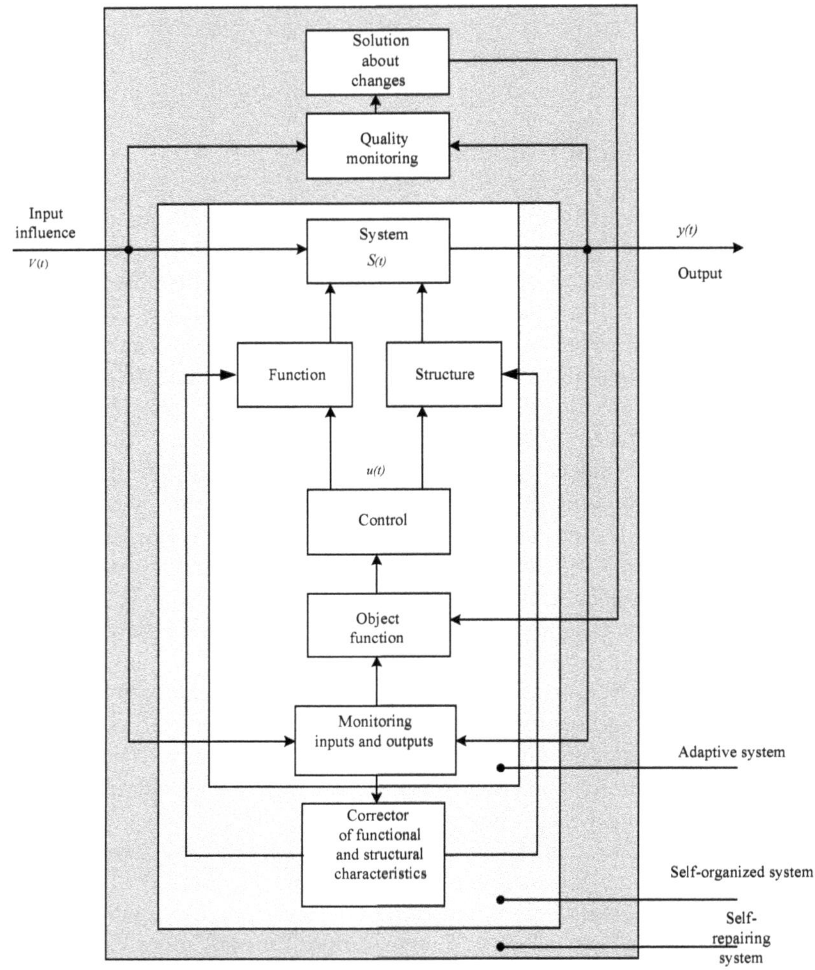

Fig. 8.11 Block diagram of self-repaired system

Recommended Literature

1. Dorf, R., Bishop, R.: Modern Control Systems. Prentice Hall PTR, Englewood Cliffs (2003)
2. Lyons, R.: Understanding Digital Signal Processing. Prentice-Hall, Englewood Cliffs (2004)
3. Sage, A., Melsa, J.: System Identification. Academic Press, London (1971)

Chapter 9
Methods of Neural Networks in Control of Telecommunication Systems

Abstract. The chapter is devoted to neural networks and their application into the control tasks. The connection of perceptron with the tasks of observation control is discussed. The peculiarities of functioning are considered for neural networks with different organization, as well as self-organized and self-taught networks. Few examples are given for practical solutions, such as the travelling salesman problem and Boltzmann machine.

9.1 Background of Neural Networks

The cybernetics was created by Norbert Weiner and his colleagues in the 1940s. It was determined as a science about control and communication in the animal and machine. John von Neumann was the first who found some analogues between processing elements of a computer and neurons.

Next was a successful attempt of formalizing the elements of intellect used in the wildlife. After applying mathematical tools, the cybernetics was born as an independent science. The neural networks (NN) play a significant role in cybernetics; they are models of parallel and distributed computing. These models are based on setting topologies and weighted connections among neurons. The specific features of these weighted computing in application to the tasks of data processing and control made the neural networks as an independent scientific discipline. In the same time, the methods of applying NN have many mutual features with the classical tasks of estimation and control discussed in previous chapters of this book.

The base element of NN is a neuron. It is an artificial element having more than one input named dendrites. Each from N inputs of a dendrite has its own weight w_i, $i = 1,2,\ldots,N$, as well as one output named an axon (Fig. 9.1). This neuron is named a perceptron of Mac Callock – Pitts. It is assumed here that the input data can be represented in analogue form, whereas the weight coefficients w_i determine the importance of this or that input signal x_i. These coefficients present in the each input channel.

V. Popovskij et al.: Control and Adaptation in Telecommunication Systems, LNEE 94, pp. 113–130, 2011.
springerlink.com

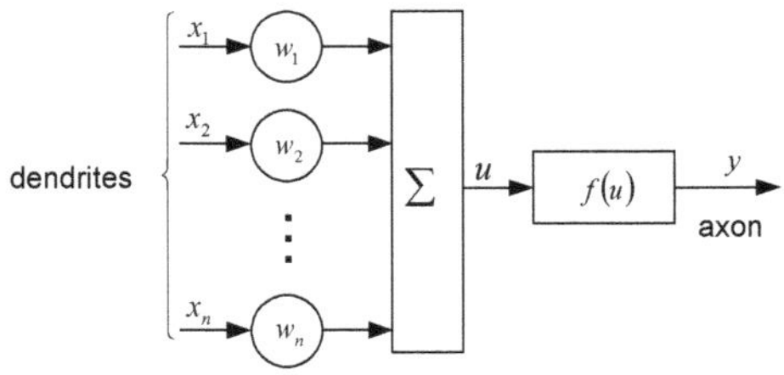

Fig. 9.1 Block diagram of neuron of Mac Callock–Pitts

The output signal of the adder is determined as:

$$u = \sum_{i=1}^{N} w_i x_i. \tag{9.1}$$

The equation (9.1) can be viewed as a weighted sum of input influences. It is quite possible that only single signal x_i possesses the weight $w_i = 1$, whereas the rest can take some arbitrary values from the interval $w_i \in [0...1]$, $i \neq j$. It is clear that the procedure (9.1) represents the degenerated transformation from the point of view of dimension. It executes the mapping of the n - dimensional space X into the single-dimensional space U. Thus, the weighting $w_i x_i$ together with the degenerated transformation allows getting these or those useful properties in the single-dimensional space U.

The adder's output is connected with the nonlinear threshold element $f(u)$. As a result, the neuron produces the following function:

$$y = f\left(\sum_{i=1}^{N} w_i x_i\right), where f(u) = \begin{cases} 0\, if\, \mathrm{u} < 0, \\ 1\, if\, \mathrm{u} \geq 0. \end{cases} \tag{9.2}$$

It is possible to have some different values for the values of thresholds. For example, the following ones formulae can be used:

$$f(u) = \begin{cases} 1\, if\, \mathrm{u} \geq 0, \\ -1\, if\, \mathrm{u} < 0; \end{cases} \tag{9.3}$$

$$f(u) = \begin{cases} 1\, if\, \mathrm{u} \geq 0, \\ -1\, if\, \mathrm{u} < 0, \\ \mathrm{u}\, if\, |\mathrm{u}| \leq 1. \end{cases} \tag{9.4}$$

Obviously, the choice of these or those values for threshold functions $f(u)$ determines the position of corresponding N-dimensional hyperplane dividing the state space by two (in the case of (8.3)) or three (in the case of (8.4)) half-spaces.

Three different diagrams of separating functions are shown in Fig.8.2. Here the function (9.2) is shown in Fig. 9.2a; the function (9.3) is shown in Fig. 9.2b; the function (9.4) is shown in Fig. 9.2c.

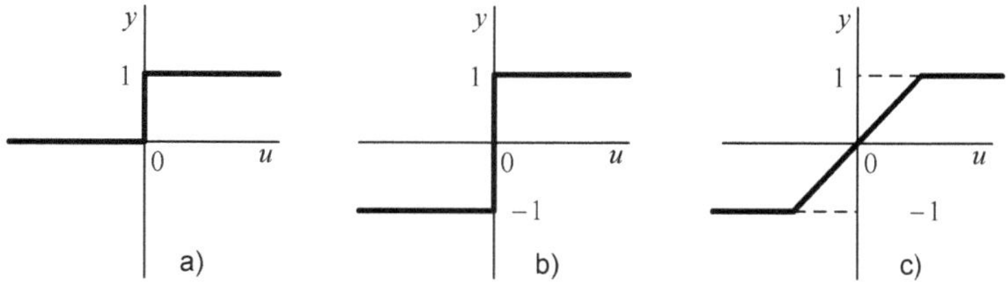

Fig. 9.2 Diagrams of separating functions $f(u)$

The sigmoid neuron has a structure similar to the neuron of Mac Callock-Pitts. But in the case of sigmoid neuron, there is a possibility of differentiability of the function $f(u)$. The differentiability allows application of efficient gradient methods of stochastic approximation (as well as Kalman-Busy filter) to solution of the control tasks. The signal function $f(u)$ is represented in the unipolar form:

$$f(u) = 1/(1+\exp\{-\beta u\}) > 0. \tag{9.5}$$

In (9.5) the symbol β means the slope coefficient of the diagram of function $f(u)$ (Fig. 9.3a).

The bipolar nature of a function is represented by the following formula:

$$f(u) = th\left(\frac{au}{2}\right) = \frac{1-\exp(-au)}{1+\exp(-au)} > 0. \tag{9.6}$$

In (9.6) the symbol a means the slope coefficient of the diagram of function $f(u)$ (Fig. 9.3b).

Fig. 9.3 Diagrams of separating functions (9.5) (a) and (9.6) (b)

Obviously, if there is $\beta \to \infty$, then the characteristic (9.5) tends to the threshold unipolar function (9.2), whereas if there is $a \to \infty$, then the characteristic (9.6) tends to (9.3).

9.2 Relationship of Neural Networks with Tasks of Observation Control

It is mentioned in Chapter 6, that the task of observation control is reduced to use of the mutual adder for weighting signals, interferences and noises received by N antennas. Next the weights w_i are controlled to minimize the interferences. As a result, such algorithms are determined as the algorithms of adaptive antenna arrays, the adaptive interference compensators and so on. Comparison of the Mac Callock-Pitts perceptron with the structure of AAA shows that these structures are very close.

The simplified block diagram of AAA is shown in Fig. 8.4. Comparison of the structure of perceptron (Fig. 8.1) and the structure of AAA (Fig. 9.4) shows that the AAA does not include the computation circuit with the function $f(u)$. Lack of this block in AAA is explained by the main goal of AAA. This goal is the suppressing of interferences on the adder output and formation of a desired structure of observation $y(t)$. It is clear that connecting some computing circuit (nonlinear element) with the output of AAA leads to some generalized algorithm. This algorithm represents the typical structure of a digital receiving device, in which some nonlinear separating function is used for determining a threshold in correspondence with a chosen optimality criterion.

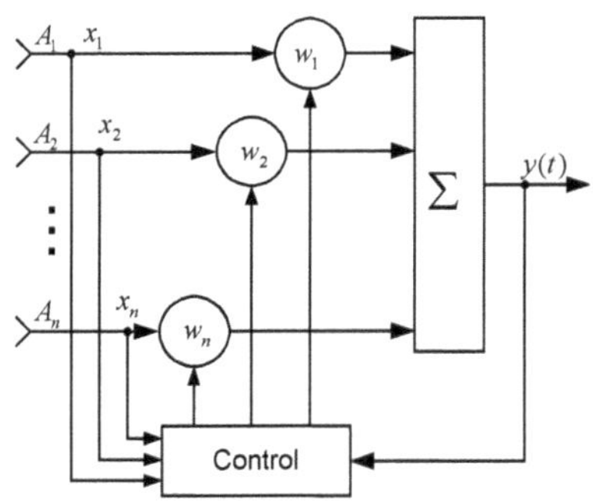

Fig. 9.4 Block diagram of adaptive antenna

In the theory of neural networks such a phenomena as the adaptive linear neuron (ALN) is known. It can be also shown that ALN implements a structure which is similar to the structure of AAA implemented using the Robbins-Monro procedure:

$$w_i(k+1) = w_i(k) + \alpha \left[y_m - \sum_{i=1}^{N} w_i(k) x_i(k) \right] x_i(k). \tag{9.7}$$

In (9.7) the following symbols have the following meanings: y_m stands for the model signal, α is a step constant, where $\alpha = (0 \div 1)$. In the same time, the neuron and AAA have more significant differences besides the lack of the decision making circuit $f(u)$. These differences are the following:

1. Input signals $x_i(k)$ of AAA are high-frequency coherent; they can be viewed as mutual copies having a phase shift due to differences in the arriving times for different antenna elements (see Fig. 7.6). If control is executed using the weight vector (WV), then AAA provides the mutual compensation of values of interference $n(k)$ on the output of the mutual adder. To do it, the algorithm (7.15) is used. The weighting coefficients w_i of AAA should be either complex (for control of amplitude and phase) or can be represented in quadratures:

$$w(t)\cos(\omega t-\varphi)=w(t)\cos\varphi\cos\omega t+w\sin\varphi\sin\omega t=w_c(t)\cos\omega t+w_s(t)\sin\omega t. \quad (9.8)$$

In (9.8) the variables $w_c(t), w_s(t)$ are the quadrature components of WV.

In contrast, the coherence is not supposed to be present in the input circuits of a nheurous. Moreover, the most often it is supposed that inputs x_i can be some constants (for example, 0 and 1) or some slowly varying functions. It should be pointed that these functions can have harmonic complex nature.

2. The adaptive antenna array is a completed purposeful object. In the same time, the neuron is not only a secluded element, but it is mostly included into some more complex system (network) having a lot of neurons. The classical neural network is a set of interrelated neurons. It permits creating a huge variety of different systems having different functions and properties.

9.3 Classification of Neural Networks and Peculiarities of Their Operation

Interrelated neurons can form different multilayer structures (Fig. 9.5). Each layer can include different amount of neurons. If the opposite is not agreed, then each output signal y_{ij} from the layer j can be connected with inputs of all neurons from the layer $(j+1)$.

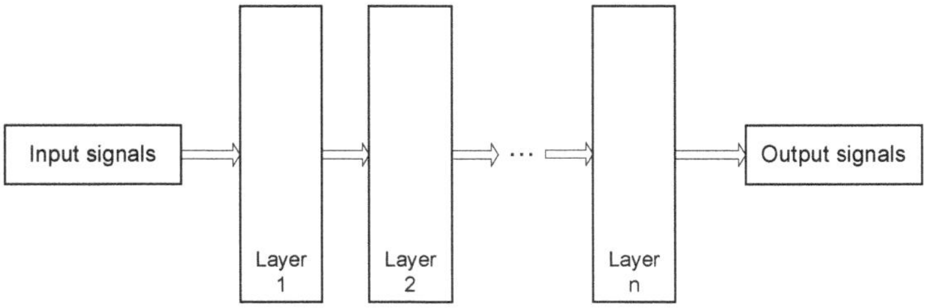

Fig. 9.5 Structure of multilayer neural network

Under the standard approach of entering for input signals, all neurons from the layer 1 receive each input signal. The layer 1 is named an input layer, whereas the layer n is an output layer. All other layers are named hidden layers.

Each layer (the output layer is an exception) can be divided by two blocks, driving (exciting) and inhibitory. The connections between the layers can be either driving (with positive weights, w_i) and inhibitory (with negative weights, $-w_i$). In the case of driving connections, any output signal of a block is represented by some monotonically non-decreasing function of a signal from a previous block. If there are inhibitory connections, then this signal is a monotonically non-increasing function.

The neuronal networks can be divided by the following constructions in dependence on the structures of communications: 1. Fully connected NN, where each neuron transmits its output signal x_i to all other neurones, including this very neuron. In such a network the input adder of a neuron is divided by two adders. The first of them calculates some linear function of input signals. The second calculates some nonlinear function of output signals generated by other neurones during the previous step of operation. The fully connected NN is also named the Hopfield network.

2. Multilayer- fully connected NN, where there are such layers that each of them is the fully connected network.

3. Recurrent NN, where the layers form a circle. In this case the last layer transmits its output signals to the first layer. Be turning on one time, these networks can operate infinitely.

4. Fully connected multilayer networks having the properties of two previous classes of the networks.

5. Radial networks (or networks with the radial function), where neurons implement some specific functions. These functions are changed radially around some appointed center and they are not equal to zero only near this center. These functions are determined as $\varphi(x) = \varphi(\|x-c\|)$. Let us name them radial basic functions. In these networks the role of a neuron is reduced to representation of the radial space around either some point used as a center or some group of points forming a cluster. The superposition of signals from all neurons allows mapping for the whole multidimensional space.

All actions taking place in our life can be interpreted and estimated according with their outcome (result). Similarly, the outcome of NN operating can be interpreted by their output signals. Let us discuss some typical operations used in NN and leading to rational outcomes (or to desired form of output signals).

The scaling is a natural operation used for processing of output signals. Standard neural networks (having no scale) are formed in such a manner that their output signals lay into some intervals, for example either interval $[-1,1]$ or interval $[0,1]$. If it is necessary to get a signal from the interval $[a,b]$, then it is necessary to transform the output signal $y \in [-1,1]$. The transformation formula is the following one:

$$y = (a+b)/2 + (b-a)y/2. \tag{9.9}$$

The rule of interpretation "a winner takes it all" is wide spread in the classification tasks. It means that the number of neurons is equal to the number of classes and the number of neurous with the maximum output signal is interpreted as the class number. Unfortunately, if there are a lot of classes, then such a simple method is too wasteful (it consumes too many output neurones).

A sign interpretation requires only $k = \lceil \log_2 m \rceil$ neurones, where m is the number of classes; in this formula the brackets are used as the sign of the ceiling function. It is constructed in the following way. Let the variables $y_1, \ldots, y_k$ correspond to output signals of neurons. Let us replace the positive numbers by ones, and negative by zeros. The obtained sequence of ones and zeros (a string) is treated as the class number in the binary notation.

An ordinal interpretation is more capacious than the sign interpretation. Using the ordinal interpretation, it is possible to use only k neurons to describe the belonging to $k!$ classes. In the case of sign interpretation usage of k neurones describes belonging to 2^kclasses. Let $y_1, \ldots, y_k$ be the output signals. Let us make their sorting and let us denote through n_i the number of the signal i after sorting process. Let the value 1 correspond to the least signal and value k to the biggest signal. Let us treat the permutation $\sigma = (n_1, n_2, \ldots, n_k)$ as a word used for encoding the class number. Obviously, it is possible $k!$ different variants of permutation. This interpreter can be used if the characteristic error of an output signal is less than $1/k$. For the case k=10 we can get the resulting accuracy less than $1/10$ and there are 10! identified classes.

As well as other systems, the neural networks can operate either on the discrete time or continuously.

In the case of the networks operating in the discrete time, the states of all neurons are equal and there are no output signals in the initial instant of time. In the zero instant of time, the input signals enter the network to determine its activity. Further, the input signals can be present in any cycle of operating. The output signals can be monitored in each cycle, too. After k cycles of operation the operating cycle is completed and the system returns in its initial state. Now it is ready to start a new operating cycle (act). The learning acts can be inserted between the operating acts. In the general case, if an operating act includes k cycles, then the network receives k sets of input signals and produces k sets of output signals. This output sequence is treated as the response of the network on the given input sequence. The simplified version is used the most often when the input signals are generated only in the initial instant of time. In this variant, the output signals are checked only in the last operating cycle, in the end of the operating act.

In the cases of the multilayer and multilayer-fully connected NN, the layers can execute different tasks. It reminds a pipelining process, but in contrast the tasks executed by the same layer can be different. The networks with the cyclic nature of operation are similar to computers. It means that each question meets corresponding response. But the mode of operation is quite different in the case of networks operating in continuous time. The mode of uninterrupted operation of NN rather corresponds to existing believes about the behaviour of animals (or human beings). The experience shows that the best results of adaptation can be reached due to alteration of the operating and learning cycles. To operate in the continuous mode, the networks with cycles are necessary. There are three types of cyclic NN, namely fully-connected, layer-cyclic and multilayer fully-connected.

9.4 Solution of Practical Problems Using Neural Networks

Let us discus some ways of using neural networks for solution of practical problems. Let us start from the well-known travelling salesman problem (TSP). Let us discuss the TSP for n towns. There are known the distances d_{XY} between each pair of towns X,Y. A travelling salesman leaves one of these towns, next he should visit each from $n-1$ remained towns. Each town is visited only once; after the trip a salesman should come back in the starting point of his travel. It is necessary to find such the order of visiting which minimizes the total distance covered by a salesman. In the case of design of networks, the solution of this task permits minimizing losses due to finding optimal routes. Let us use the Hopfield network for finding a solution of TSP.

Let the Hopfield network include $N=n^2$ neurons and let the state of neurones be described by double subscripts v_{Xi}. Let the subscript X determine the town name, whereas the subscript i determines a position of the town in the salesman's route. Let us form the function of computing energy needed to solve the TSP. In this function the state with the least energy corresponds to the shortest route. The energy function should satisfy the some requirements.

Firstly, it should support the stable state in the following matrix form

$$V=\{v_{Xi}\}. \tag{9.10}$$

In (9.10) the rows correspond to towns, whereas the columns correspond to their numbers in a particular route. Each row and each column contain only one 1, all other elements are equal to zero.

Secondly, the energy function should support only such solutions represented by (9.10) which correspond to short routes.

The energy function shown below satisfies these requirements. It is represented as the following one:

$$E=(A/2)\sum_X\sum_i\sum_{j\neq i}v_{Xi}v_{Xj}+(B/2)\sum_X\sum_i\sum_{Y\neq X}v_{Xi}v_{Xj}+(C/2)(\sum_X\sum_i v_{Xi}-n)^2+$$
$$+(D/2)\sum_X\sum_{X\neq Y}\sum_i d_{XY}v_{Xi}(v_{Y,i+1}+v_{Y,i-1}). \tag{9.11}$$

In (9.11) three first members support the first requirement, whereas the fourth member provides supporting second requirement. The first member is equal to zero, if each row X includes not more than one 1. The third member is equal to zero, if the matrix includes exactly n symbols 1. The short routes are supported by the fourth member. Its subscripts i are taken by modulo n to show that the town number n is a neighbour of the town number $(n-1)$ in the route number i. It means that the following equality $v_{Y,n+j}=v_{Y,j}$ takes place. As a number, the forth member is equal to the route length. The canonical expression of the computing energy function can be represented as the following one:

$$E=-(1/2)\sum_X\sum_i\sum_Y\sum_j W_{Xi,Yj}v_{Xi}v_{Xj}-\sum_{xi}I_{Xi}v_{Xi}. \tag{9.12}$$

Using expressions (9.11) and (9.12), the weights of the Hopfield network can be obtained. They are the following:

$$W_{Xi,Yj} = -A\delta_{XY}(1-\delta_{ij}) - B\delta_{ij}(1-\delta_{XY}) - C - Dd_{XY}(\delta_{j,i+1}+\delta_{j,i-1}), I_{Xi} = Cn.$$

In this expression, the symbol δ is the Kronecker delta (Kronecker symbol).

The simulation of the Hopfield network shows that the solution with the best quality is given by the network whose neurons have the sigmoid characteristic. If the neurons have step transitions, then such a network comes to the final states with routes having the quality not far from the random routes. Numerous researches show that the solution quality of the minimization task for the energy function (9.11) depends significantly on the choice of the production sigmoid unipolar function of neuron activation in the neighbourhood of zero. If the value of derivative is small, then the minimums of energy are placed in the center of the solution hypercube. If the value of derivative is rather large, then the Hopfield network will occupy the hypercube node corresponding to the local minimum of energy. Besides, the choice of coefficients A,B,C,D influences the final quality of solution. Nowadays, the investigation of methods of the optimal choice for those coefficients is a subject of very thorough research.

Now let us discuss the Boltzmann machine (MB) used for solution of combinatorial optimization tasks. The mathematical base used by the Boltzmann machine is an algorithm simulating the process of solidification of either liquids or melts (it is an algorithm of anneal imitation). It is based on the ideas from two different areas, namely the statistical physics and combinatorial optimization. This task can be also interpreted for assignments of tasks in a distributed telecommunication (or computing) network. The MB can implement this algorithm as parallel and asynchronous. The MB is represented by a quadruple $B = (N,E,W,V_0)$, where N is the number of neurons, $E = \{(i,j)\}$ is a set of connections among the neurons, all autoconnections belong to this set ($(i,i) \in E$). Each neuron can be in one from two states, namely either 0 or 1. The state V_k of BM is determined by the states of neurones $V_k = (v_1^k,\ldots,v_N^k)$, in the initial instant of time the BM is in the initial state V_0. Each connection (i,j) has its weight w_{ij} represented by some real number. These connections form a set W. The connection (i,j) is called active in the state V_k, if there is $v_i^k v_j^k = 1$. The weight of connection (i,j) is interpreted as some quantitative metric of desirability for given connection to be active. If there is $w_{ij} \gg 0$, then the activity is very desirable. If there is $w_{ij} \ll 0$, then the activity is very undesirable. As it is in the Hopfield model, the connections in the BM are symmetric ($w_{ij} = w_{ji}$).

The concept of consensus is introduced for the state V_k of BM. It is determined by the following formula:

$$C_k = \sum_{i,j} w_{ij} v_i^k v_j^k. \tag{9.13}$$

The consensus C_k is interpreted as some quantitative metric of desirability that all connections (i,j) are active in the state V_k. For example, this metric is suitable for the tasks of design of telecommunication systems. In this case the consensus is used as a profit function. Each connection is accounted in this sum only once. The set of neighbours $V^{(k)}$ is formed for each state V_k. The neighbour state $V_{k(i)} \in V^{(k)}$ can be formed from V_k for changing the state of the neuron number i. It can be expressed in the following way:

$$V_j^{k(i)} = \begin{cases} v_j^k \; if \; j \neq i, \\ 1 - v_j^k \; if \; j = i. \end{cases}$$

The difference between the consensuses V_k and $V_{k(i)}$ is determined as

$$\Delta C_{kk(i)} = C_{k(i)} - C_k = (1 - 2v_i^k)\Big(\sum_{(i,j) \in E(i)} w_{ij} v_i^k + w_{ii} \Big). \tag{9.14}$$

In (9.14) the set $E(i)$ is a set of connections for the neuron i. It is obviously that the values of $\Delta C_{kk(i)}$ can be calculated at the same time for all sets $V_{k(i)} \in V^{(k)}$.

Let us discuss the maximization of consensus. The interstate transition of BM with maximization of the consensus is executed due to the following stepwise procedure. Each its step includes two parts. The first part is connected with generation of the neighbour set $V_{k(i)}$ for each state V_k. Next, it is estimated whether the state $V_{k(i)}$ can be taken as the next state of BM. If it is possible, then the test result is equal $V_{k(i)}$, otherwise to V_k. The state $V_{k(i)}$ is accepted with the probability determined in the following way:

$$P_{kk(i)}(t) = 1 / \left[1 + \exp(\Delta C_{kk(i)}/t) \right]. \tag{9.15}$$

In (9.15) the symbol $t \geq 0$ is a controlling parameter (it can be either time, or cost, or temperature).

The process of maximization of consensus starts from the high value t_0 of the parameter t and some initial state V_0 chosen in the arbitrary way. During the process the value t is decreasing from t_0 to zero. When the value of t advances zero, the states of neurons are stable and the total Boltzmann machine is stabilized in its final state. In practice, the BM stabilizes in a state corresponding to some local maximum of consensus. This local maximum is close (or equal) to the global maximum.

The initial state for each neuron i is determined by the following expression:

$$t_0^{(i)} = \sum_{(i,j) \in E(i)} |w_{ij}| + |w_{ii}|.$$

The reduction rule for the parameter t can be expressed in the following way:

$$t_{j+1}^{(i)} = \alpha t_j^{(i)}.$$

In this rule the parameter α is some positive number. The following parameters should be chosen for this procedure. The first of them is the number of tests L which are conducted without change of parameter t (L is some function of N). The second is the number M of consecutive tests which do not change the state of BM (M is some function of N). This number is used as a criterion of the process completing.

Now let us discuss the application of BM for solution of the TSP. There is a general approach for programming combinatorial tasks to be solved using the BM. In the case of BM, each solution is represented by a set $\{x_1, \ldots, x_N\}$, where

$x_i \in \{0,1\}$, N is the number of neurons in the network, and x_i is a neuron state. There are two rules used for the choice of connections and weights in the network.

The first rule (rule $R1$): all local maximums of the consensus function correspond to acceptable solutions of a task. The second rule (rule $R2$) the better the acceptable solution is, the greater is the consensus for corresponding state of the BM. Let us paraphrase the TSP for the case of BM. Now these rules are the following ones.

The rule $R1$: the state of BM corresponds to a local minimum of consensus function if and only if this state corresponds to an acceptable route. The rule $R2$: the shorter is the route, the higher is the consensus for corresponding state of BM.

Each neuron corresponds to one element of the matrix $n \times n$, the states of neurons are denoted as v_{Xi} (where n is the number of towns to be visited). The consensus function is determined as

$$C_k = \sum_{(Xi,Yj)} w_{Xi,Yj} v_{Xi}^k v_{Yj}^k.$$

The set of connections in the network is determined as the union of the three disjoint subsets. The first of them is the subset E_d determining the set of connections carrying the information about the distances between the towns (between the nodes of the network). It can be represented in the following way:

$$E_d = \{(Xi,Yj) | (X \neq Y) \wedge (i = (j+1) mod n)\}.$$

The second subset is the set E_i determining inhibitor connection in the network. It can be represented as the following one:

$$E_i = \{(Xi,Yj) | (i \neq j) \wedge (X = Y) \vee (i = j) \wedge (X \neq Y)\}.$$

The last component of the set of connections is the set E_b determining the shift connections. It is represented as:

$$E_b = \{(Xi,Yj) | (X = Y) \wedge (i = j)\}.$$

In these subsets the sets $X, Y, i, j = 1, \ldots, n$ are used. The total number of connections is equal to $2n^3 - n^2$.

The inhibitor connections guarantee that, eventually, it will be no more than only one 1 in each column and each row of the matrix. The shift connections guarantee that each row and each column will contain at least one 1. Therefore, the connections E_i and E_b guarantee fulfilment of restrictions in the tasks; their weights make equal contributions into consensuses of all acceptable routes.

The connection $(Xi,Yj) \in E_d$ is active only in the case if a route include the direct path form the town Xinto the town Y. The weight of connection $(Xi,Yj) \in E_d$ is equal to the distance between the towns X and Y taken with the negative sign. Therefore, the negative contribution of the connection E_d is proportional for a given route to the length of the distance. Thus, the maximization of the consensus function corresponds to minimization of the route length.

It is proven that the requirements $R1$ and $R2$ take places for the consensus C_k, only in the case if the weights of connections are chosen in the following way:

$$\forall (Xi,Yj) \in E_d : w_{Xi,Yj} = -d_{XY},$$
$$\forall (Xi,Yj) \in E_i : w_{Xi,Yj} < -\min(\mu_X, \mu_Y),$$
$$\forall (Xi,Yj) \in E_b : w_{Xi,Yj} > \mu_X,$$

where there is $\mu_X = \max\{d_{XP} + d_{XQ} | P, Q = 1, \ldots, n \wedge (P \neq Q)\}$.

To make investigations, the authors used the following parameters $d = 0.95$, $L = 10$, and $M = 100$. There were conducted 100 tests with $n = 10$ and 25 tests for $n = 30$ using the different initial states of the BM. In the case of $n = 10$ the obtained solution differed from the optimum solution on 14% (it was worse than the optimum). The probabilistic nature of BM operation allows obtaining slightly better results in comparison with the results produced by the Hopfield model.

9.5 Recurrent Neural Networks on the Base of Perceptron

The operating process of a distributed telecommunication system can be viewed as a recurrent process. Its representation as a neural network allows obtaining some important results. The multilayer recurrent networks correspond to further evolution of unidirectional perceptron-based networks by introducing corresponding feed-backs. The feed-back can come from either output or hidden layer of neurons. Each feed-back loop includes the element of unit delay giving to the information flow the property of unidirectionality. It means that the output signal from the previous cycle of operation is treated as known a priori. This signal just increases the length of the network input vector. The recurrent network represented in such a way operates as a unidirectional perceptron network. It should use a learning algorithm for adaptation of values of synaptic weights. The algorithm is rather complex due to existed dependence of current signals (signals in the instant of time t) from their values in previous cycles. It results in a really cumbersome formula used for calculating the gradient vector.

Let us reduce our discussion of recurrent networks based on the output multilayer perceptron by the most popular models of networks such as recurrent multilayer perceptron (RMLP), RCMI Translation Research Network (RTRN), and Elman network. Let us start from the perceptron network with feed-back connection.

The simplest way for constructing the recurrent network on the base of unidirectional neural network is introduction of a feed-back into the perceptron network. The RMPL-network is formed as a result of this introducing. The structure of RMLP is shown in Fig. 9.6. In this structure, the elements of unit delays are denoted by symbols z.

The RMPL-network is a dynamic network which is characterized by delays of input and output signals. These signals are combined into the network input vector.

Let us discuss the system having only one input node $x(k)$, one output neuron, and one hidden layer. Such a system implements the following mapping:

$$y(k+1) = f(x(k), x(k-1), \ldots, x(k-(N-1)), y(k-1), \ldots, y(k-P)). \quad (9.16)$$

In (9.16) the quantity $N-1$ corresponds to the number of delays of the input signal, whereas symbol P denotes the amount of neurons in the hidden layer. In this case the RMLP-network can be characterized by the following triplet of numbers (N, P, K). The vector x entered the network input as the following:

$$x(k) = [1, x(k), x(k-1), \ldots, x(k-(N-1)),\ \ y(k-P), y(k-P+1), \ldots, y(k-1)]^T. \quad (9.17)$$

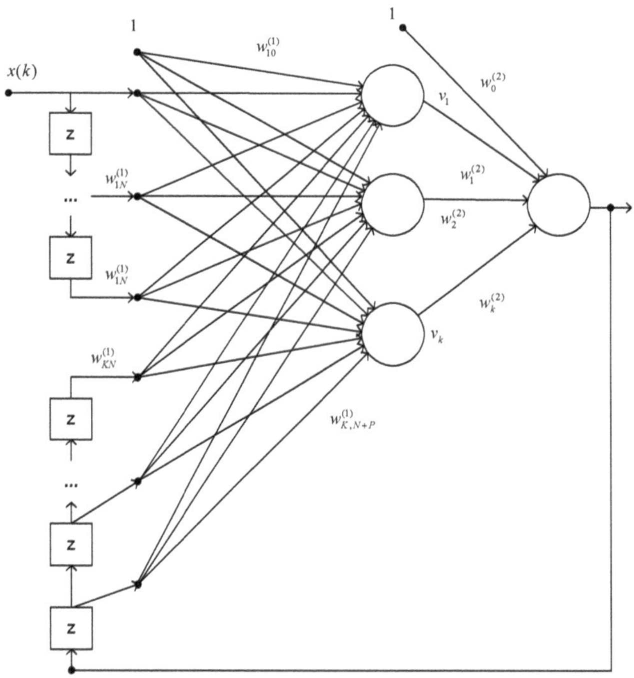

Fig. 9.6 Structure of RMLP-network

Let us assume that all neurons possess the sigmoid activation function. Let us use the symbol u_i to denote the weighted sum of signals for the neuron number i from the hidden layer. Let the symbol g denote the weighted sum of signals for the output neuron. Now the output signals of the neurons are represented by the following dependencies:

$$u_i = \sum_{j=0}^{N+P} w_{ij}^{(1)} x_j; v_i = f(u_i); g = \sum_{i=0}^{K} w_i^{(2)} v_i; y = f(g).$$

The RMLP-network can be successfully implemented for simulating dynamic processes in the on-line mode. For example, it can be used for the imitation of nonlinear dynamic objects. In this case the RMLP-network is used as a model of a given process, whereas the algorithm of weight refining is used as the identification procedure for parameters of this model (Fig. 9.6). Comparison of control algorithms discussed in Chapter 6 and the algorithm represented by Fig. 8.7 leads to conclusion about their generality. The misalignment $e(k)$ in this model is used for control of RMPL, whereas the function $y(k)$ is used as a model signal.

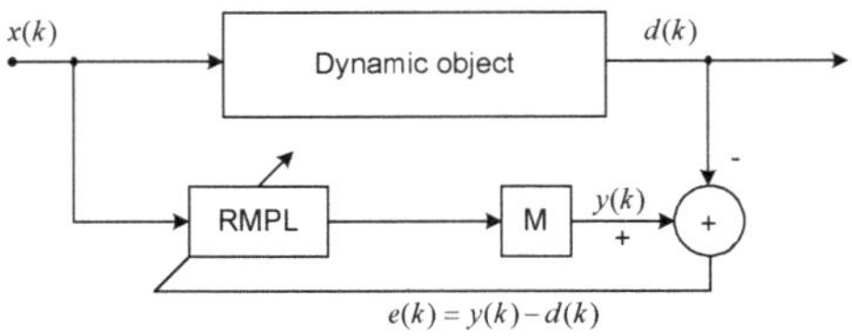

Fig. 9.7 Usage of RMPL-network for solution of the identification tasks

This system operates in the following manner. It is made comparison of the output signal $y(k)$ of the model with the output signal $d(k)$ of the dynamic object. The value of misalignment (error) $e(k) = y(k) - d(k)$ is calculated. The misalignment is used for controlling the process used for refining the parameters of the neural network. The system includes a block M used for scaling the output signal $y(k)$ in such a manner that its dynamic level occupies the same diapason as the level of the dynamic object $d(k)$. The scaling process is achieved due to appropriated choice of the gain coefficient of the block M.

9.6 Self-organization and Self-learning of Neural Networks

The learning process of the neural network with self-organization is organized on the base of the competition among the neurons. The main goal of such a network is such an ordering of neurons which minimizes the value of the expected distortion. The ordering of neurons is an appropriate choice their weights. The expected distortion is estimated by the approximation error of the input vector **x** by the weight values of the winner among the neurons. This error is called the quantizing error. If there are p input vectors **x** and if the Euclidean metrics is applied, then the quantizing error can be represented as the following one:

$$E = (1/p) \sum_{i=1}^{p} \left\| x^i - w_{win} \right\|^2 . \tag{9.18}$$

In (9.18) the symbol w_{win} stands for the weight of the winner neuron in the appearance of the vector x^i.

Such an approach is called the vector quantization (VQ) or clustering. If the vectors x^i enter the input of network one by one, then the numbers of the winners form so called coding table. In the classical case of the solution of the coding the algorithm of K-averaging is applied. It is called the generalized Lloyd algorithm.

In the case of the neural networks, the algorithm WTA (winner takes all) is an analogue of the Lloyd algorithm. According to the WTA algorithm, the activity of each neuron is calculated after entering of each vector **x**. The winner of competition is a neuron having the strongest output signal. It means that it possesses the maximum value of the scalar product $(\mathbf{x}, \mathbf{w})$. It can be shown that it is the same as the minimal Euclidean distance between the least input vector and the weight vector of the neurons (if the normalized vectors are used). The winner gets the right for refining its weights in the direction of the vector **x** using the following rule:

$$w_{win} \leftarrow w_{win} + \alpha(x - w_{win}). \tag{9.19}$$

In (9.19) the symbol α is a learning coefficient. The weights of other neurons are not refined. The algorithm permits taking into account the "fatigue" of neurons by calculation of the number of victories. The elements having the least activity are stimulated to make their chances higher. As a rule, this modification is applied on the starting stage of learning process; it is terminated after activation of all neurons. This learning mode is implemented as the mode called CWTA (conscience winner takes all). It is considered as one of the fastest and best algorithms of self-organization.

In the WTA algorithms, only one neuron can be learned. There is the WTM (Winner Takes Most) algorithm which is widely used for the learning of self-organized networks. In the WTM algorithm, the neurons from the nearest neighbourhood of the winner refine their weights, too. In this case, the degree of refining depends on the distance between the winner and the given neighbour. The refining process can be defined as the following general dependence

$$w_i \leftarrow w_i + \alpha G(i,x)[x - w_i]. \tag{9.20}$$

The formula (9.20) is applied for all neighbours of a winner. Let I be the number of a winner. The classical WTA algorithm corresponds to the case when the function $G(i,x)$ is determined as the following one:

$$G(i,x) = \{1 \; for \, \mathrm{i} = \mathrm{I}, 0 \; for \, \mathrm{i} \neq \mathrm{I}\}. \tag{9.21}$$

Obviously, the algorithms (9.19) - (9.20) belong to the class of recursive algorithms of stochastic approximation discussed in details in Chapter 6. There are a lot of variants of WT algorithm, distinguished by the form of the function $G(i,x)$. For example, let us discuss the classical Kohonen's algorithm.

The Kohonen's algorithm is one of the earliest learning algorithms used in the networks with the self-organization based on competition. Because of it, there are different versions of this algorithm. In the classical Kohonen's algorithm, the network is initialized by assignment of specific space positions for its neurons and connecting them with their neighbours on the continuing basis. Such a network is called

SOFM (self-organizing feature map). When a winner is chosen, then its weight is refined as well as the weights of its nearest neighbours. Therefore, the winner is undergone by adaptation, as well as its neighbours. In the classical Kohonen's algorithm, the function of neighbouring $G(i,x)$ is determined as the following one:

$$G(i,x) = \{1\ for\ d(\mathrm{i},\mathrm{I}) \leq \mathrm{L},\ 0\ for\ d(\mathrm{i},\mathrm{I}) > L\}. \tag{9.22}$$

In (9.22) the symbol $d(\mathrm{i},\mathrm{I})$ stands for the Euclidean distance between the weight vectors of the neuron-winner (it has the number I) and the neuron i. The coefficient L shows the level of neighbouring; its importance is decreased up to zero during the learning process. The neighbouring of such kind is called rectangular.

The Gaussian neighbouring is another kind of neighbouring applied in the Kohonen's algorithm. In this case, the function $G(i,x)$ is represented as the following one:

$$G(i,x) = \exp(-d^2(i,x)/2\lambda^2). \tag{9.23}$$

The level of adaptation is determined for the neurons-neighbours using two factors. The Euclidean distance between the winner-neuron (having number I) and the neuron number i is the first factor. The second factor is the level of neighbouring, λ. In the rectangular neighbouring, each neuron from the neighbouring of winner is adapted in the equal degree. In the case of the Gaussian neighbouring, the adaptation level is different and it depends on the value of the Gaussian function. As a rule, the Gaussian neighbouring provides the better learning results and better organization of the network than the rectangular neighbouring.

The self-organizing feature map goes through two stages of learning. The first stage is connected with ranking (ordering) of its elements. It is executed in such a manner that it reflects the space of the input elements. The second stage is devoted to refinement of their positions. As a rule, the process is represented visually by usage of two-dimensional data and constructing the corresponding space. For example, the input vectors are chosen in the arbitrary way on the base of the homogeneous distribution in the some square. Next, the learning of the map starts. In some specific instances of time the image of the map is created. The map is similar to the network structure shown in Fig. 9.6. The network elements are connected by lines to show their relative location. In the beginning, the map looks rather "crumpled", but it is unfolding and smoothing out gradually during the learning process. The final result of the learning process is a map reflected the whole input space; this map is rather regular (its elements are distributed practically evenly). Let us discuss the example of the map represented as a square having 49 elements. The learning process is conducted for 250 sources of data. The initial distribution of cluster elements in the center of input space is shown in Fig. 9.8. The learning process is shown in Fig. 9.9, whereas Fig. 9.10 shows the distribution reached near the end of the learning process.

The initial distribution is determined by the random set of weight values. Next, it is unfolded step by step. The final stage is shown in Fig. 9.10. The elements of the map are ordered. The regularity of the map will increase after the end of the final

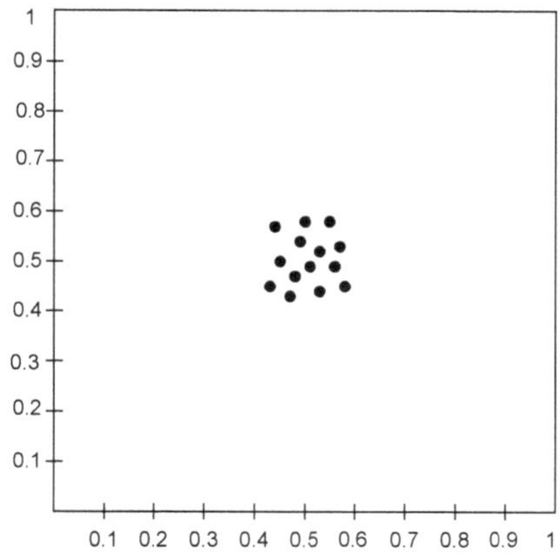

Fig. 9.8 Weight vectors after initialization by random values for the range 0.4 – 0.6

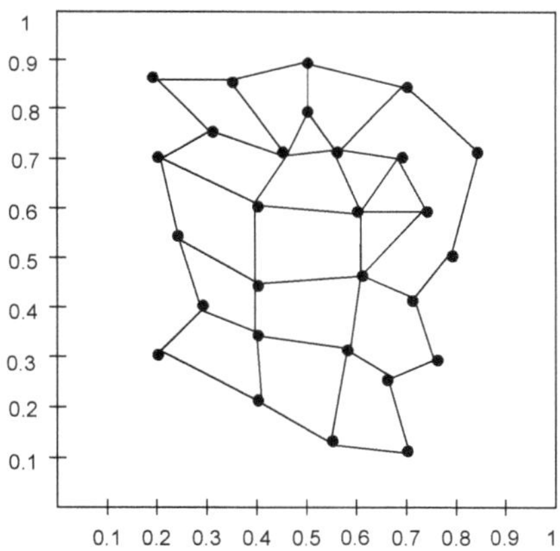

Fig. 9.9 The map after 20 itterations

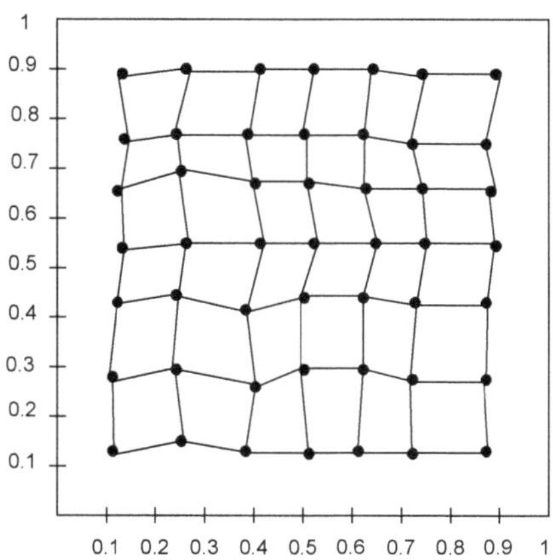

Fig. 9.10 The map near the end of learning process

phase of convergence process. Let us point out that the result of learning depends strongly on the learning data and choice of learning parameters. It is true for the networks having other types of organization, too.

Recommended Literature

1. Bishop, C.: Neural Networks for Pattern Recognition. University Press (1995)
2. Carling, A.: Introducing Neural Networks. Sigma Press (1992)
3. Fausett, L.: Fundamentals of Neural Networks. Prentice-Hall, Englewood Cliffs (1994)
4. Haykin, S.: Neural Networks: A Comprehensive Foundation. Macmillan Publishing, Basingstoke (1994)
5. Patterson, D.: Artificial Neural Networks. Prentice-Hall, Englewood Cliffs (1996)
6. Ripley, B.D.: Pattern Recognition and Neural Networks. Cambridge University Press, Cambridge (1996)

Chapter 10
Methods of Automata Theory in Telecommunication Systems

Abstract. The chapter is devoted to discussion of the theory and methods of multi-functional control automata. The methods of Petri nets and E-nets are also discussed. These methods are used for simulation, analysis and development of telecommunication networks.

10.1 Short Introduction into Automata Theory

A lot of models are developed describing automata used for controlling some objects. An automaton belongs to the class of specialized cybernetics objects characterized by the matrix of transitions. This matrix describes relations between input signals X and output signals Y. The specific feature of automata is existence of internal states. An abstract automaton can be determined by the following quintuple:

$$A = (S, X, Y, \delta, \lambda). \tag{10.1}$$

In (10.1) the symbol S stands for the finite set of internal states; symbols X, Y determine finite sets of calibrated input and output signals (variables) forming input and output alphabets; the transition function $\delta : S \times X \to S$ determines the next state of automaton; the output function $\lambda : S \times X \to Y$ determines the current output of automaton. The model 10.1 is called finite state machine (FSM).

Thus, the FSM is a cybernetic system determined by the sets of inputs and outputs and characterized in the space of states. At the same time, the FSM differs from the classical dynamic system. The main difference is the specialization of FSM. To compare these models, let us show the differential model of the cybernetic system. The comprehensive characteristic of such a system is its state equation represented as the following one:

$$x(k+1) = F(k+1,k)x(k) + G(k+1,k)\xi(k). \tag{10.2}$$

V. Popovskij et al.: Control and Adaptation in Telecommunication Systems, LNEE 94, pp. 131–152, 2011.
springerlink.com

In (10.2) the matrices $F(\cdot)$ and $G(\cdot)$ determine correspondingly the inertial properties and the level of functional states of the system.

In contrast, both the influences $X$and responses $Y$are not arbitrary in the case of FSM (10.1). These issues are represented as some sequences of calibrated pulses. The inertance of FSM is also standardized; it is determined by the digitalization frequency of input pulses.

The comparison of properties of the FSM (10.1) and the cybernetic system (10.2) shows that the FSM is a system with restricted properties. It targets some narrow class of specific influences. But such a specialization allows expansion for other properties. Particularly, it makes possible the development of adequate models describing interrelations among digital elements. Let us point out that the digital technology is the basis of modern telecommunication systems. The application of FSM is especially successful in the tasks of analysis of structural and functional properties of TCS.

Obviously, automata can execute control functions in a system. In this case they are called control automata. The outputs Y of a control automaton present the control influences entering either some network element or other control automaton. At the same time, the FSM can be treated as a controlled element, because it operates under influence of some input influences X. So, these both terms (control FSM and controlled FSM) are used according to the tasks solved by a particular FSM.

10.2 Models of Control Automata

A control automaton can be considered as a device implementing some algorithm of functioning determining some sequence of execution of specific operations or control procedures target some object. It is worth pointing that the models of control automata permit analysis not only functional, but some specific structural properties of complex interrelated systems. These models successfully combine abilities for investigation of given properties.

During its operation, a control automaton generates some sequence of control signals entering some controlled object. This sequence is determined by the operating algorithm. It depends on both current states of FSM and external influences, which can be formed by other FSM or by a human being. Therefore, the interconnection of an FSM and an object is the base of the system "a control automaton – a controlled object".

In the case of TCS the interconnection "agent – manager" exists. In this case, the source of message (a manager) is viewed as a control automaton, whereas the receiver of this message (an agent) represents a controlled object. The control automaton is represented by the transport station sending some signals; the receiving station (terminal) represents the controlled object. At the same time, if the method of spatial-temporal encoding is used in the adaptive communication channel, then the automatic choice of emission parameters is executed to overcome the phenomenon of multipath propagation. In this case the receiving station should change the parameters of the transport station. It means that the receiving station controls the transport station. Therefore, the transport station is a controlled object, whereas the receiving station is a control automaton.

A control automaton can be represented as a mathematical model of a device with finite memory; this device is used for transforming discrete information. The automaton can be characterized as a device with input and output channels; in any instance of time it is in some internal state.

Modern progress tendencies of telecommunication systems are such, that the systems more and more acquire properties of an automaton distributed in both space and time. This automaton operates using some digital alphabet and it has the finite set of internal states. The state control is executed in accordance with protocols, user requests and restrictions of resources.

The formalization of the model of control automaton is executed in the following way. In each cycle t the automaton executes the following actions. Its input channel receives input signals x represented by the letters of its input alphabet X. At the same time, the output signals y are generated by its output channel. These signals are the letters of the output alphabet Y; each output signal depends on an internal state s from the state alphabet S and on input influence (a letter) x. The internal state s' in the next cycle $(t+1)$ is determined by the current state s and the current input letter x. The input data are transformed into some output data. The transformation law is determined by the functions δ and λ:

$$y = \delta\ (a, s), \forall y \in Y; \forall x \in X; \forall s \in S; s' = \lambda\ (x, s), \forall s' \in S. \tag{10.3}$$

The transformation of words represented in the alphabet is the main characteristic of an FSM. This property can be determined by the sets $(S, X, Y, \delta, \lambda)$ of a given control automaton.

Some specific signals, as well as arbitrary enough influences can be treated as the words and letters of alphabets X and Y. These signals and influences form the sets of alphabets and states S. Particularly, the letters can be represented by predicates which are formalized logic-mathematical objects determined for some set having n elements. The formalized language used for predicate logic is represented by an alphabet having four groups of symbols. These groups are: the predicate variables, the predicate constants, logic operations (conjunction, disjunction, implication, equivalence, negation, existential quantifier, and universal quantifier), and auxiliary symbols (brackets, commas).

The operation of FSM can be described by the following system of recurrent equations:

$$\begin{cases} s(t+1) = \delta(s(t), x(t)), \\ y(t) = \lambda(s(t), x(t)). \end{cases} \tag{10.4}$$

The block diagram of FSM corresponding to (10.4) is shown in Fig. 10.1.

The transition function δ is represented by the state diagram or the transition graph. Some example of state diagram is shown in Fig. 10.1. It is a weighted one-way graph whose vertices (S_0, S_1, S_2) correspond to the states and arcs correspond to inter-state transitions. The weights of arcs show the symbols causing a particular transition. If a transition is caused by more than one input signal, then all these

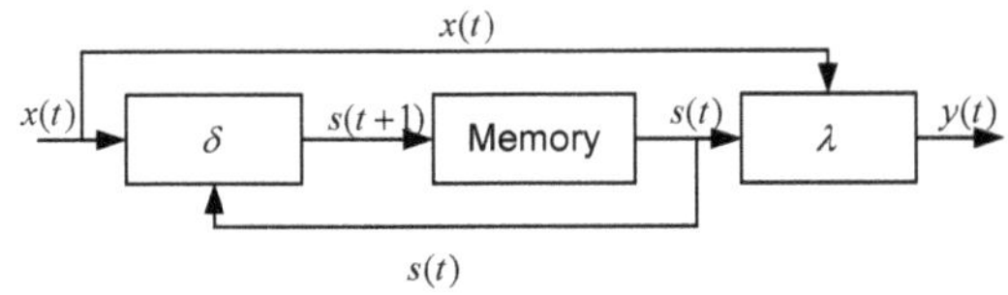

Fig. 10.1 Block diagram of FSM

signals should be shown above the particular arc. The arcs are marked by the pairs j/k, where j represents input data and k represents output data.

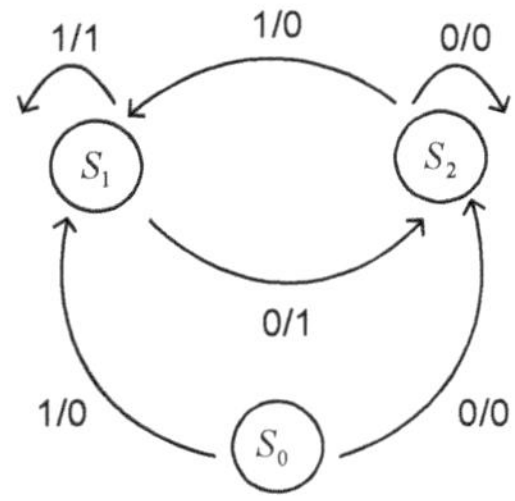

Fig. 10.2 State diagram of FSM

The matrix or tabular representations are also convenient for representing the transition function δ. In such a matrix each row corresponds to one state and each column corresponds to one input signal. In the cell of the matrix an output signal is written. It corresponds to some actions executed by an FSM in the situation corresponding to these state and input symbol.

Let us discuss the classification of control automata. There are three main groups of FSM.

The first group includes finite and infinite automata. The automata are infinite if some of their alphabets are infinite. It could be either alphabet X, or alphabet S, or alphabet Y. It is the most general class of automata. Such a generality leads to decrease of their practical meaning. The infinite automata do not find a wide practical application. But the finite FSMs having final sets X, S, Y find very wide application for solution of practical problems.

The second group includes indeterministic (stochastic, probabilistic) automata having some arbitrary relations or random functions instead of deterministic functions δ and λ. As a rule, these automata operate in the asynchronous mode.

The third group consists of automata with variable structure (or the variable-structure automata). They are the finite automata $(S \times S, X, Y, \delta, \lambda)$ having two inputs where there is fixed some infinite sequence α (mega-word) in the alphabet S. The first input of such an FSM receives arbitrary words in the alphabet S, whereas the second input receives the beginning of the sequence α having the same length (as the words from the first input). Due to such organization some restrictions are introduced for the set of pairs of input words.

Apart from these three main groups, there are other automata belonging to the class of finite FSM but having their own specific features. The following subclasses can be found among the finite FSM:

1. If the transition and output functions of the finite automaton are replaced by some fuzzy relations, then the fuzzy automata are obtained.
2. If each output letter is determined only by an input letter, then the automata without memory are obtained. They also are called functional elements or combinational circuits.
3. If each initial letter for any initial state is determined by some restricted part of an input word, then the finite-memory automata are obtained.

Against the number of functions and data transformations executed by automata they can be divided by two classes. The first of them includes automata with fixed structure (or monofunctional automata). The second class consists on reconfigurable (or multifunctional) automata.

The monofunctional FSM has a rigid (fixed) structure providing execution only single data transformation $\{X\} \to \{Y\}$. For such an FSM, the value of its functionality is equal to 1 ($L = 1$).

A controlled automaton is called multifunctional (M-automaton) if it can implement some set of automaton transformations$\{X_i\} \to \{Y_i\}$, using the tuning set $D_A = \{A_i\}i = \overline{1,L}$, where $L > 1$. There are two ways for implementing M-automaton. It can be implemented using a set of separate interrelated fixed-tuned (having no reconfiguration) mono-automata with reconfigurable structure. The second way is representation of M-automaton as a single multi-input multi-output block having a reconfigurable internal function.

An automaton is called reconfigurable if there is a set of automaton transformations implemented by it and a tuning algorithm exists providing tuning on implementation of each from these transformations. According to the tuning principle, the M-automata are divided by three classes: with functional, structural and programmed tuning.

The functional tuning is such a tuning when a tuning code Z_i is unchangeable during the whole time of execution of transformation A_I. In this case the connections among the functional elements of M-automaton can be either fixed or reconfigurable depending on the transformation to be executed. In the last case, the changing connections inside the circuit of M-automaton is equivalent to transforming its internal structure. If the tuning process is connected with change of interconnections inside the structure of automaton, then such a tuning is called structural.

The reconfigurable automaton can be represented as a set of automata having mutual inputs and outputs. A tuning determines the automaton whose outputs are treated as outputs of the total automaton. Thus, the reconfigurable automaton cannot implement anything except of automaton transformations.

If the values of outputs for any tuning are determined only by the input signals, then such an automaton is called a combinational reconfigurable automaton or a multifunctional logic module. If the output values depend on the automaton states (maybe, only for some tunings), such an automaton is called a reconfigurable automaton with memory.

In the case of programmed reconfigurable automaton, the transformations A_I are executed for some amount of steps $n_{\tau i}$, when the tuning code is changed for each step of transformation.

The functional tuning is the simplest kind of tuning; it is characteristic for the simplest M-automata (such as Mealy FSM, Moore FSM, or trivial FSM). The structural tuning is more complex; it is characteristic for the automata with complex organization. These automata are represented by composition of some functional elements (automata) and elements (automata) of commutation or connection. The tuning code for such M-automaton includes two parts. The first part is responsible for the tuning of functional automata. The second part is used for the tuning of commutation automata. At the same time, the functional and commutation (switching) automata can have either functional or structural mode of tuning.

Let us discuss two basic types of finite synchronous automata, namely the Mealy and Moore FSM models.

The Mealy FSM is represented by the following systems of functions:

$$\begin{cases} \chi(t+1) = \delta\left[\rho(t), \chi(t)\right], \\ \beta(t) = [\rho(t), \chi(t). \end{cases} \tag{10.5}$$

In (10.5) the symbol $p(t)$ denotes the state of FSM inputs in the cycle t; these input states form a set $\rho = \{\rho_1, \ldots, \rho_N\}$. The symbol $\beta(t)$ stands for the output state of FSM in the cycle t; they form a set $\Lambda = \{\lambda_1, \ldots, \lambda_k\}$. The symbol $\chi(t)$ stands for the internal state of FSM in the cycle t. The internal states form the set of states:

$$\chi = \{\chi_0, \chi_1, \ldots, \chi_{S-1}\}. \tag{10.6}$$

Let us introduce the definition for the complete state of FSM. The complete state μ of FSM in the cycle t is determined by its internal state and input state in this very instant of time. The internal state in the instant of time $(t+1)$ is determined by its state in the instant of time t. This dependence is represented as the following one:

$$\chi(t+1) = \varphi[\mu(t)]. \tag{10.7}$$

The expression (10.7) shows that the output state of FSM depends on its internal state and input state. Using the complete states, this statement can be expressed by the following formula:

$$\beta(t) = [\mu(t)]. \tag{10.8}$$

The mapping among different pairs of “the inputs $\rho(t) \in P$, the internal states $\chi(t) \in \chi$” and values of transition states $\chi(t)$ and outputs $\beta(t)$ is determined by functions of transitions and outputs, correspondingly.

In the Moore FSM, the transition function is determined by expressions (10.5) or (10.7). The output function of Moore FSM differs from the output function of Mealy FSM. It is expressed by the following formula:

$$\beta(t) =' [\chi(t)], \tag{10.9}$$

It means that the outputs of Moore FSM are determined only by its states. So, there is no direct dependence among the inputs and outputs of Moore FSM.

Despite the absence of the direct dependence among its inputs and outputs, the Moore FSM can be viewed as a particular case of Mealy FSM. Indeed, using the following dependence $\chi(t) = \delta[\chi(t-1), \rho(t-1)]$, one can get the following final expression:

$$\beta(t) = \lambda[\chi(t-1), \rho(t-1)]. \tag{10.10}$$

It follows from (10.10) that the output state in the current clock time depends on the input state, but from the input in the previous clock time. This formula shows the main difference between the Mealy and Moore FSM. In the case of Mealy FSM, the output state appears simultaneously with the input state. In the case of Moore FSM this appearance is delayed by one cycle.

The operation of asynchronous automaton is determined by the following pair of equations:

$$\chi(t+1) = \delta[\rho(t+1), \chi(t)], \tag{10.11}$$

$$\beta(t+1) = \lambda[\rho(t+1), \chi(t+1)]. \tag{10.12}$$

The equation (10.11) shows that the internal state for the instant of time $(t+1)$ is determined by the input state in the same instant of time and its internal state in the previous instant of time. It means that the input state demanding the transition in some internal state is kept unchangeable till the fulfilment of required transition. On the other words, the internal state in the current instant of time depends on its input state in the same instant of time.

The output state of asynchronous FSM is determined by (10.12). This formula is the same as for synchronous Mealy FSM. Of course, this dependence can be replaced by some formula where the outputs depend only on internal states (as it is for the Moore FSM).

To analyze the multielement control automata with simultaneous operation of some interrelated elements, the Petri nets can be used. The Petri net (PN) is a mathematical model of discrete dynamic systems (including the informational systems). This model allows the qualitative analysis of discrete dynamic systems (detecting lockouts, critical situations and bottlenecks of the systems).

10.3 Models of Control Systems Based on Petri Nets

The Petri nets are the finite automata consisting of some standard elements, such as places (positions), transitions, direct arcs and tokens (marks). These elements are show in Fig. 10.3.

In the graphical interpretation of a Petri net, the arcs can connect only the nodes having different types. For example, a place can be connected with a transition, or a transition can be connected with a place.

There are two basic conceptions allowing constructing the model of operation of some dynamic system as a Petri net. These conceptions are events and conditions. The basic PN is represented by the vector $N = (P, T, A, M_0)$. This vector has the

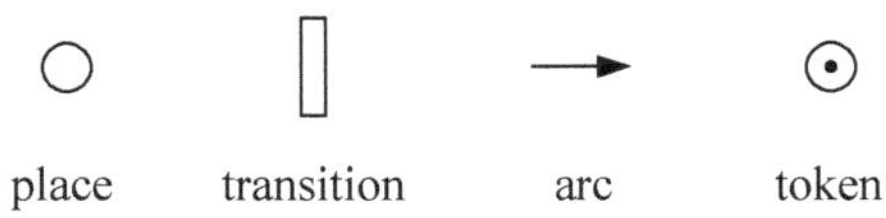

Fig. 10.3 Graphical symbols used for elements of basic Petri nets

following elements: the set of places $P = \{p\}$; the set of transitions $T = \{t\}$; the mapping A representing the arcs and their multiplication factor (ratio) and determined as $P \times T \bigcup T \times P \rightarrow N_0$, where $N_0 = N \bigcup \{0\}$; the mapping μ representing the distribution of tokens along the places of network (marking of PN) and determined as $P \rightarrow N_0$. The graphical representation of PN forms a direct two-partite graph where places are shown by circles (or ellipses) and the transitions are represented by either a rectangle or bar (barrier).

One of possible PN is shown in Fig. 10.3.

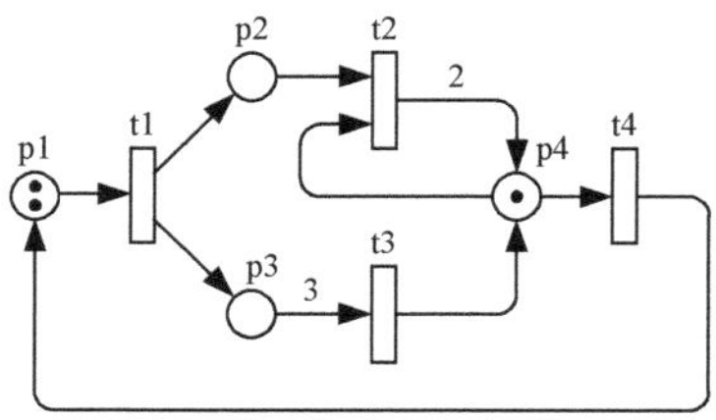

Fig. 10.4 Example of graphical representation of Petri net

The places of PN represent some conditions. If a condition is true, then the corresponding place is replaced by a token. Transitions represent some events taking place into a particular PN. Events are the actions (processes) run in the simulated dynamic system. Appearance of an event corresponds to execution of a corresponding transition. The marking of a PN determines the state of the simulated system. The execution of transitions changes the marking. Obviously, it corresponds to the changing the system state due to implementing some event. The change of marking, in turn, leads to possibility of execution for new transitions. It means that now new events are possible in the new system state. For example, the set of conditions can be simulated for simulation the process of packet transmission with confirmation of the place. This set includes the following conditions: the presence of the packet in the output buffer of a particular device, existence (or absence) of interferences (or congestion) into the channel in use, existence of either the information packet or confirmation into the input buffer of a particular device, or existence (or absence) of errors in the packet. In this case, the transitions can describe such events as the successful transmission of the packet, or its loss, or its damage during the transmission process.

The graphical representation of PN is convenient; it gives the visual tools for entering and editing both simple and hierarchical PN. But it requires usage of some complex specialized software. Nowadays, there are other approaches for representation of Petri nets. The following of them can be mentioned: a matrix representation, an algebraic representation, and a representation on the base of basic fragments.

The algebraic (set-theoretical) representation of a PN is the following one:

$$N = (G, M_0). \tag{10.13}$$

In (10.13) the following components are presented: the graph $G = (P, T, A)$ is a graph of a PN; the set $P = \{p\}$ is a finite set of places; the set $T = \{t\}$ is a finite set of transitions; the set A is a finite set of arcs.

In the case of the PN shown in Fig. 10.4 these sets are the following: $A = \{(p_1,t_1,1),(p_2,t_2,1),(p_3,t_3,3),(p_4,t_4,1),(p_4,t_2,1),(t_1,p_2,1),(t_1,p_3,1),(t_2,p_4,2),$ $(t_3,p_4,1),(t_4,p_1,1)\}$,$P = \{p_1,p_2,p_3,p_4\}$,$T = \{t_1,t_2,t_3,t_4\}$ and the set $\mu_0 = \{(p_1,2),$ $(p_4,1)\}$ is an initial marking of the PN. To the sake of simplicity, only the nonzero vales of the functions are shown in these sets.

The matrix representation of a PN is the following one:

$$N = (B, D, \bar{\mu}_0). \tag{10.14}$$

In 10.14 there are $B = \|b_{i,j}\|$, $b_{i,j} = A(p_i,t_j)$, $i = \overline{1,m}$, $j = \overline{1,n}$, $m = |P|$, $n = |T|$, $D = \|d_{i,j}\|$, $b_{i,j} = A(t_j,p_i)$, $i = \overline{1,m}$, $j = \overline{1,n}$, $m = |P|$, $n = |T|$,

The matrix representation of the PN shown in Fig. 10.4 is shown below:

$$B = \begin{bmatrix} 1 & 0 & 0 & 0 \\ 0 & 1 & 0 & 0 \\ 0 & 0 & 3 & 0 \\ 0 & 1 & 0 & 1 \end{bmatrix}; \quad D = \begin{bmatrix} 0 & 1 & 1 & 0 \\ 0 & 0 & 0 & 2 \\ 0 & 0 & 0 & 1 \\ 1 & 0 & 0 & 0 \end{bmatrix}.$$

The initial marking is represented by the vector $M_0 = (2\ 0\ 0\ 1)$.

If there are no loops in a PN, then the network can be represented as the following one:

$$N = (C, \bar{M}_0). \tag{10.15}$$

In our example, the matrix

$$C = D - B = \begin{bmatrix} 1 & -1 & -1 & 0 \\ 0 & 1 & 0 & -2 \\ 0 & 0 & 3 & -1 \\ -1 & 1 & 0 & 1 \end{bmatrix}.$$

The matrix approach has the following advantages. It permits a very simple description of a PN-model. The second advantage is an existence of software tools using namely the matrix representation of Petri nets. But this approach is not free from some drawbacks. One of them is the most important: this form is very bulky in the cases of detailed representation of a PN-model.

The algebraic representation of a PN-model is free from the drawbacks of the matrix approach. But it is complex enough and requires some specific skills achieved by training. The approach based on usage of the basic fragments leads to very compact and simple in use representation. But it develops some inconveniences in research of hierarchical Petri nets.

The following interpretation of the transition execution is assumed in the Petri nets: a transition can take place only if there is at least one token in each of input places. When a transition is executed, it deletes from its input place the number of tokens equal to the weight of corresponding arc. This very number is placed into the output place of this transition. If a PN includes a transition which can be executed, it will be obligatory executed. If more than one transition is ready, then these transitions are executed one by one. But there is no preliminary deterministic knowledge about the sequence of their execution. Thus, the operating of a PN can be treated as a random sequence of discrete events.

The rules used in Petri nets for executing transitions allow visually representation for control procedures. Both control procedures based on the principles of Ponselle (control by influence) and Watt (control by deviation) can be represented. The simplified model of an uncontrolled system is represented as the Petri net and it is shown in Fig. 10.6.

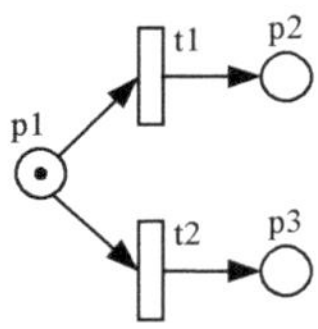

Fig. 10.5 Simplified model of an uncontrolled system based on PN

The shown initial marking leads to the random character of the simulated system's behaviour. It means that there is no preliminary knowledge about the order of execution for transitions t_1 and t_2. If this system is supplemented by two control places (p_4 and p_5), then the Ponselle based-control can be implemented in the simulated system. The modified model of the system with the control by influence is shown in Fig. 10.6.

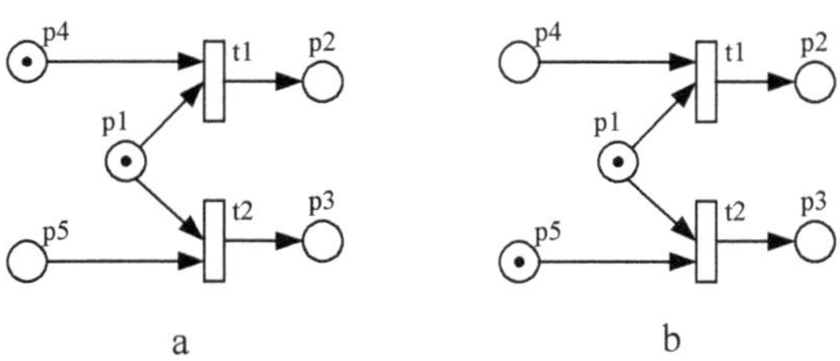

Fig. 10.6 Model of system with control by influence based on PN

Alterations of the initial marking for given PN allow control by the system behaviour. If the initial marking is executed in the manner shown in Fig. 10.6a, then the transition t_1 is always executed as the first transition. If the initial marking is executed in the manner shown in Fig. 10.6b, then the transition t_2 is always executed as the first transition, whereas the transition t_1 is forbidden.

The model of a system with control by deviation is shown in Fig. 10.7. This model is based on the PN with feedbacks. Due to the feedbacks, the transitions t_1 and t_2 are executed in turn.

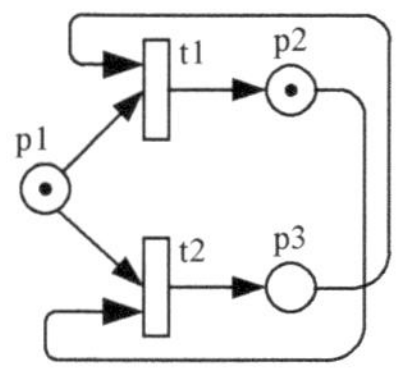

Fig. 10.7 Model of system with control by deviation based on PN

A very wide class of different systems can be simulated using the apparatus of Petri nets. This class also includes the control systems of TCS, as well as the objects controlled by TCS. But there is even more important task which can be solved using Petri nets. It is a task of analysis allowing the understanding of the behaviour of a simulated system. One of the important peculiarities of Petri nets is a possibility of existence (or absence) of deadlocks. The deadlocks arise in the real systems due to distribution of limited resources among the interrelated processes. The deadlocks are the subjects of many research made in the theory of TCS. In the case of PN-models, the deadlocks correspond to the deadlock markings. Exposure of the deadlock markings is connected with the analysis of the Petri net's liveness. If it is necessary to find either possibility or its lack for some state of a simulated system, then they use procedures for analysis of reachability of Petri nets.

There are a lot of methods allowing solution of above mentioned problems. Nowadays, the following methods are widely used for analysis of Petri nets: construction of the tree of reachable markings (the reachability graph); investigation of the structure of a PN-graph; the invariant analysis.

The reachability graph gives the most complete characteristic of the behaviour of a PN. It is possible because the graph counts all possible markings and sequences of executions in a particular PN. This approach allows conducting analysis of such issues as the liveliness, restrictions, reachability for the given marking. As a rule, existed algorithms and programs constructing the reachability graph operate with PN having not more than 1000 nodes.

The methods analysing the structure of PN are known for some limited class of Petri nets. But this methods are more economical in terms of both time and memory consumptions in comparison with the graph methods. They have the polynomial complexity with the third degree for the practical Petri nets.

Methods of invariant analysis are connected with the search of basic sets of solutions for some matrix equations. There are different software tools for implementing these methods. The tools are based on the methods of the linear algebra; they can be applied to networks having not more than 1000 nodes.

In the majority of cases, the complexity of analysis algorithms depends exponentially on the number of the nodes of a PN. Because of it, tremendous attention is paid to approaches allowing decrease of algorithmic complexity. Two such approaches are known.

The first of them is reduction. It determines the order of transformations for a particular PN leading to decrease for the number of nodes with preservation of analysed properties. There are some practical algorithms and software tools allowing efficient reduction of Petri nets.

The second approach is the decomposition. In this case, the initial network is divided by some fragments. The analysis is conducted independently for each fragment. Next, each fragment is replaced by a single place (or transition) and the behaviour of the final hierarchical PN is conducted. This approach can be applied if the components of decompositions are determined in the process of constructing Petri net. Moreover, it should be proven that the components keep the properties of the total resulting PN.

In the practical discrete systems, the following situation is quite general. There are two devices which are ready to act, but one of them has priority for launching. The classical PN cannot simulate this situation. But there is a modification of PN, where each transition has its priority. Such PN are called the prioritised Petri nets; in such nets each transition has its priority. The firing rules for transitions are supplemented by the following: if more than one transition can fire, then higher-priority transition will fire. The class of prioritised Petri nets has the same power as the class of Turing machines.

Special inhibitory arcs can be introduced in classical Petri nets. These arcs check the zero marking. If a transition fires, then the marking is changed from nonzero to zero value. Such extension of classical PN is called an inhibitory network. The class of inhibitory Petri nets has the same power as the class of Turing machines.

There are practical systems where probabilities of transitions among the states of the subsystems depend on the current state of the total system. The system preventing the congestion of router is an example of similar subsystem. The classical PN do not allow investigation of the systems where there are probabilistic intercommunicated processes. To overcome this limitation, the stochastic Petri nets are created.

All tokens of the classical PN are Boolean; it means there is no difference among them. One of the most popular extensions of the classical Petri net is so called coloured Petri net (CPN). It is possible to use very complex token in the case of CPN. Sometimes, the type of a token is called its colour. Such an approach allows obtaining more compact models in comparison with the corresponding basic PN-model. It is connected with the fact that in CPN each place can simulate a variety of conditions.

It is presumed in the theory of PN that any transition fires immediately (without any delay). But in the practical systems the overwhelming majority of events need some finite time interval to be occurred. It is impossible to neglect the existence of delays occurred in execution some operations. The temporal Petri nets are used for simulating systems where some timing characteristics should be taken into account. The apparatus of temporal PN can be used for investigation many operating properties of the real telecommunication systems (such properties as network delay, system throughout, and so on).

The hierarchical Petri nets allow simulating really complex systems using the modular principle. In this case, the simulated systems can be described using either top-down or bottom-up approaches. The existed modules (components) can be used more than once; besides, new components can be created using these typical modules.

There are some other extensions of classical Petri nets taking into account the specifics of a particular application domain. Such extensions can be pointed out as E-networks, PROT-networks, predicative networks, and so on.

10.4 Models of Control Systems Based on E-Nets

The E-networks can be viewed as finite state machines whose structures can be represented by networks. They have more detailed internal structure. Because of it, they can be viewed as powerful tools for adequate representation of telecommunication networks. The additional advantage of E-nets is their powerful abilities for simulation. In this area they excel considerably the capabilities of FSM and they are close to abstract machines.

Formally, the E-net is represented by a bipartite oriented graph. This graph is represented as the following one:

$$E = (P, H, L, D, A, M_0). \tag{10.16}$$

Following components can be found in (10.16): set P is a finite set of places with subsets B (a finite set of peripheral places) and R (a finite set of decisive places). H is a finite set of transitions including the sets ; L is a direct incidence function D is an inverse incidence function; A is a finite set of transition characteristics including pairs $\alpha = (\tau(\alpha_i), q)$, ($\tau(\alpha_i)$ is the time of firing for a given transition, q is a transition procedure); M_0 is an initial marking of a network.

As an apparatus used for simulation of telecommunication systems, the E-nets have the following possibilities:

1. They can easily simulate parallel intercommunicated asynchronous processes reflecting the general dynamic of operation for the discrete system.

2. They provide whatever semantic interpretation for their components. It permits the simultaneous simulation of both data flows and hardware. The simulation can be done for the complete TCS.

3. They allow different treatment of their components according to the level of abstraction. It provides constructing hierarchical models where a transition can be translated into some sub-network with lower abstraction layer. It significantly decreases the contradiction between the requirements of simplicity and adequacy of a model.

The representation of a control system as some graph of E-net has one particular goal. This goal is reduced to obtaining a convenient and obvious algorithm of simulation. Next, some universal simulation program can be developed on the base of this graph. It allows excluding the programming process from the simulation. The programming process is replaced by some graph accelerating the analysis process for protocols. At the same time, this approach brings down the requirements for the level of programming skills from the people developing a model.

Let us discuss the logic of transitions used in the E-nets. The logic of transitions is set up by the setting the allowed changes of markings. The firing of the transition t (Fig. 10.8) is executed if there is a token in the input place and there is no token into the output place. It can be described by the following equation: $(1,0)\overset{t}{\mapsto}(0,1)$. The transitions of this kind provide reflections for the events having place after justification of one condition.

Fig. 10.8 Representation of t-transition

The condition of enabling for transitions of this type is the following one:

$$p_1 \in L(t)\{M(p_1)=1\} \wedge p_2 \in D(t)\{M(p_2)=0\}. \qquad (10.17)$$

In (10.17) the symbol L stands for the function of direct incidence; D is the function of inverse incidence; $M(p_i)$ is a marking for the place p_i.

The branching of condition flows is simulated by F–transition (Fig. 10.9). The F–transition is used for simulating branching processes into the systems to be simulated. These processes are, for example, the broadcasting of packets, the initialization of a few parallel processes, and so on.

The F–transitions fire if there is a token in the input place and there are no tokens in the pair of output places: $(1,0,0)\overset{F}{\mapsto}(0,1,1)$.

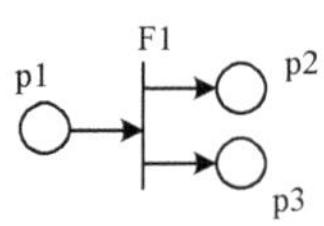

Fig. 10.9 F–transition

The condition of enabling for F–transitions is the following one:

$$p_1 \in L(F)\{M(p_1)=1\} \wedge p_2 \in D(F)\{M(p_2)=0\} \wedge p_3 \in D(F)\{M(p_3)=0\}. \qquad (10.18)$$

If some event takes place only if two conditions are true, then so called J–transition (Fig. 10.10) is used. The J–transition simulates the combining some data flows. The J–transitions fire if there are tokens in the pair of input places and there is no token in its output place: $(1,1,0)\overset{J}{\mapsto}(0,0,1)$.

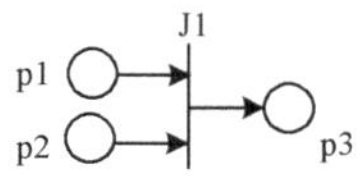

Fig. 10.10 J–transition

The condition of enabling for J–transitions is the following one:

$$p_1 \in L(J)\,\{M(p_1)=1\} \wedge p_2 \in D(J)\,\{M(p_2)=1\} \wedge p_3 \in D(F)\,\{M(p_3)=0\}\,. \tag{10.19}$$

If it is necessary to simulate control procedures, the X–transitions are used (Fig. 10.11). The main element of this transition is a controlling place r_x. Its value determines the transferring tokens into one of output places: $(0{,}1{,}0{,}0)\overset{X}{\mapsto}(0{,}0{,}1{,}0)$; $(0{,}1{,}0{,}1)\overset{X}{\mapsto}(0{,}0{,}1{,}1)$; $(1{,}1{,}0{,}0)\overset{X}{\mapsto}(0{,}0{,}0{,}1)$; $(1{,}1{,}1{,}0)\overset{X}{\mapsto}(0{,}0{,}1{,}1)$.

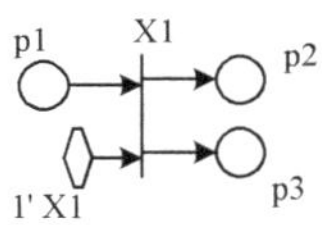

Fig. 10.11 X–transition

The condition of enabling for X–transitions is the same as for F–transitions.

To simulate the priority processing different data flows, as a rule, the Y–transition is used (Fig. 10.12). This transition is represented by the following system of expressions: $(0{,}1{,}1{,}0)\overset{Y}{\mapsto}(0{,}0{,}1{,}1)$; $(0{,}1{,}0{,}0)\overset{Y}{\mapsto}(0{,}0{,}0{,}1)$; $(1{,}1{,}1{,}0)\overset{Y}{\mapsto}(0{,}1{,}0{,}1)$;

$$(1,1,0,0)\overset{Y}{\mapsto}(0,0,0,1);(1,0,1,0)\overset{Y}{\mapsto}(0,0,0,1).$$

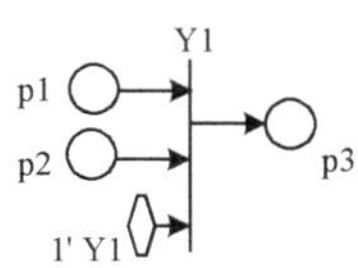

Fig. 10.12 Y–transition

The condition of enabling for Y–transitions is the same as for J–transitions.

So, five main types of transitions are introduced. Their mutual usage allows simulating different situations characteristic for practical dynamic systems. Let us discuss the application of E-nets for simulating telecommunication systems.

The sets P and H satisfy to the following conditions: $P \neq \emptyset$, $H \neq \emptyset$, $P \cap H = 0$. It means that the graph of E-net should include at least one transition and one place. Besides, a graph node cannot belong to both sets P and H simultaneously.

The functions of direct and inverse incidence (L,D) set the following rules:

$$L : B \times H \rightarrow \{0,1\}, D : H \times B \rightarrow \{0,1\}. \tag{10.20}$$

The expression (10.20) shows that these functions determine that there are no possibility for connection of the elements belonged to the same set. Besides, they describe the sets of input and output elements.

The places of the E-net show the conditions for enabling different events, represented by transitions. For example, the process of packet processing can be represented by some sequence of steps. In this case, the presence of a packet on the step (n-1) is a condition for its transition to the step n. The places of E-net represent the processing stages for packets, or states of hardware, or states of the communication channel, and so on.

The transitions of the E-net simulate some events for the level of execution for all necessary conditions. Also, they show some operations connected with the events by modification of tokens. For example, they can reflect the setting TCP connection. The set of operations and conditions of their enabling are described by the transition procedure ρ. In the general case, the operation Ξ for the descriptor number i for the token t_k can be represented in the following form (if the predicate number j is true, where this predicate reflects the set of required conditions):

$$l_{R_{ij}} = \{(M(t_R(i))) := \Xi(M(t_R(i)))\}. \tag{10.21}$$

As an intermediate conclusion, we can point out that a E-net sets both specific deterministic structure of a model and its operating algorithm.

In telecommunication systems, the necessity arises in describing the control processes, such that the control procedures depend on values of either input or output control signals, as well as on the contents of data flows by themselves. Some examples of similar tasks are the route operating, the control of commutator (switch) operating with priorities, the simulation of decision making process connected with the choice of particular mode of data transmission, and so on. The describing these processes can be either stochastic (for example, the simulation of error arising) or deterministic (sending a data packet to a corresponding receiver) character.

The apparatus of E-nets includes transitions with X and Y. Their actions are reduced to analysis of some predicate describing a transition procedure and transferring a token to the corresponding place. But sometimes in the simulation of some application domain it is not enough to use only the token parameters to obtain the value of a given predicate. Therefore, it is necessary to expand the abilities for describing these predicates.

The macro-transition with type MX allows description of such events by redirection of the token flow into one from n output places. The graphical representation of the macro-transition MX is shown in Fig. 10.13. The place number receiving a token is determined by the value of the predicate ρ from the decisive place r_i. The value of predicate can depend on such factors as the appointed attribute of a token in the input place, the marking value, the given law of probability distribution.

Let some condition of the transition enabling be true and let this condition presume the existing of a token in the input place. In this case, the predicate ρ is analysed. If a token is absent in the output place having number ρ, then this token is

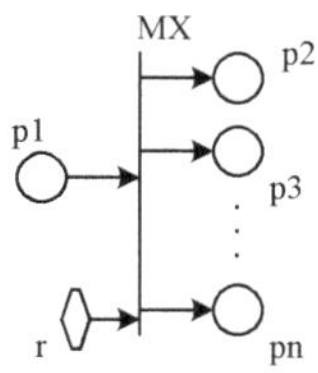

Fig. 10.13 Macro-transition of MX type

transferred from the input place into this output place. The transferring is executed after some time, determined in the vector of attributes.

The following system of equations determines the law of operation for MX-transition: (1,1,0,f,...,f)$\overset{MX}{\mapsto}$(e$_0$,0,1,f,...,f); (2,1,f,0,...,f)$\overset{MX}{\mapsto}$(e$_0$,0,f,1,...,f); .... ; (n,1,f,f,...,0)$\overset{MX}{\mapsto}$(e$_0$,0,f,f,...,1). In these equations, the symbol f stands for the marking of the place $p_i \in P$, which is equal to $M(p_i) = 1 \vee 0$, the symbol e_0 denotes not fixed marking of the decisive place after the transition's firing.

The MY-transition (Fig. 10.14) allows simulating the priority in data processing, where data enter from n directions. For example, such a transition allows simulating the hub operation with priorities of packets.

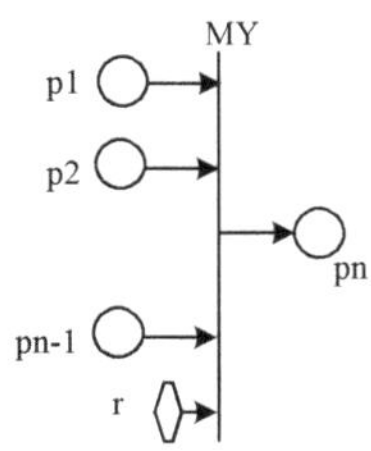

Fig. 10.14 Macro-transition of MY type

The mode of operation for the MY-transition can be described using two equations. They are: (m,f,f,...,M(p$_m$)=1,...,f,0)$\overset{MY}{\mapsto}$(e$_0$,f,..., M(p$_m$)=0,...,f,1) and (m,f,f, ...,M(p$_m$)=0, M(p$_{m+1}$)=1...,f,0) $\overset{MY}{\mapsto}$(e$_0$,f,..., M(p$_m$)=0, M(p$_{m+1}$)=0,...,f,1). These equations mean the following. If a value of the predicate in the decisive place is equal to m and if there is a token in the input place number m and if there is no token in the output place, then the token from the input place m is transferred into the output place. But if there is no token into the input place m, the token is transferred from the first input place having the number (priority) exceeded m.

Let us discuss the methods used for analysis of algorithmic properties of E-nets. The complex of analysed properties is determined by the nature of an object to be simulated. The decision about reasonability for research of some model's property can be made after clarification of this property's place into the semantic of a system to be investigated. There are the following algorithmic properties of E-nets.

The property of boundedness characterizes the capacity of conditions whose images are the places of a net. If this property is analysed, then such characteristics can be checked as the sizes of sending and/or receiving buffer units, or the performance of some devices, or the abilities of a system for efficient distribution some restricted resources (performance, throughput, and so on). The property of safety is more rigid condition. It characterizes the limitedness of the places of an E-net. A net is safe if the following condition takes place: $\forall M(p) \in E\,\{\forall p_i|M(p_i) \leq 1\}$, $i = 1...n$, $n = |P|$. It means that there is no possibility for appearing more than one token in any place of the net.

The property of conservatism presumes equality of values for the initial and all other markings of a net. It means that the amount of tokens inside a net is always the same. A net is conservative if there is a vector $K = \{k_1, ..., k_n\}$ such that for all possible markings M_i the following equality takes place:

$$\sum_{i=1}^{m} k_i M(p_i) = \sum_{i=1}^{m} k_i M_0(p_i). \tag{10.22}$$

In (10.22) the symbol M_0 denotes the initial marking of the place p_i; the symbol M stands for an arbitrary marking of the place p_i which is reachable from the initial marking; the symbol k_i shows the amount of tokens in the place number i.

If tokens are interpreted as some limited resource, then it is important to provide the safety analysis for a given system.

The property of consistency (determinacy) is determined as the equality between the number of transitions enabled by the marking M_i and the number of transitions firing for this marking. If a network's model has no its property, then processes taking place into the system are stochastic.

The property of potential liveliness presumes the reachability of enabling markings for all transitions from the initial marking. The property of liveliness presumes that all transition processes have the potential liveliness for any reachable marking. The net is called live if all its transitions are live.

The property of terminal existing is the following one. There are tokens entering the open network from external environment. The property assumes that all such tokens either leave the network or they are accumulated in some determined macroplaces (absorbents) in the process of this network's operation. This property exists for E-nets simulating some completed processes. These processes can include, for example, the processes of making a connection, or connection release, or processing of some determined number of packets and so on.

The property of stability presumes that for all acceptable markings a net inevitably transits into its initial marking (after some operating process). This property indicates the cyclic character of the processes taking places into a simulated system.

Two main methods are used for recognition of the pointed above properties of E-nets. The first of them is the method of state equations; the second is the method of covering tree.

The main idea of the method of state equations is in constructing recurrent equations. These equations link the vectors of initial marking M_{i-1} and resulting marking M_i with incidence functions and the control vector. The control vector sets the rules for enabling transitions. As a rule, this approach is used for checking the existence of boundedness and deadlock markings.

In the case of the covering tree, each marking reachable from the initial marking M_0 corresponds to some state of both E-net and an object simulated by this network. The corresponding sequence of firing transitions is represented as a word $L(E)$. The word $L(E)$ corresponds to the prehistory of this state. The sets $R(E,M_0)$ and $L(E)$ characterize completely the system behavior represented in the terminology of E-nets. Their analysis allows obtaining the solubility for practically all algorithmic problems, as well as estimates some quantitative characteristics of a system to be simulated by E-net.

As a rule, these sets ($R(E,M_0)$,$L(E)$) are represented as some directed graph. Each node of this graph corresponds to some marking; the arcs show the sequences of firing transitions. Thus, each arc corresponds to some transition between the pair of markings. In common case, the graph of reachable markings can be infinite; its form depends on a conception used for constructing a model using E-nets.

If there is the infinite set of place, then the graph is infinite, too. In this case, it is practically impossible to investigate the graph to get an answer about existence of some property. To get these answers, it is necessary to change the procedure of graph construction. The changing is reduced to the following rule: the set of nodes should include only markings M' satisfying to some condition. Two conditions are used, namely: $M' = M$ or $M' > M$, where M is one of the nodes obtained before. Such a graph is called the complete covering graph. Its analysis allows obtaining solutions for all algorithmic problems mentioned above.

10.5 Turing Machines

The automata theory includes a very important issue used in solution of practical problems. This issue is the theory of Turing machines. The Turing machine (TM) is a name used to define some abstract computing machine. Generally speaking, the TM is a mathematical model (or mathematical equivalent) of an algorithm describing operation of infinite automaton having the infinite set of internal states. The TM has the infinite external memory represented by the tape. This tape is divided by cells; each cell can include any letter from some alphabet.

In any discrete instant of time, the Turing machine stands in some state. It analyses one of the cells of its tape. The symbol from analyzed cell (it can be any letter or empty cell) is received by the TM.

Despite the fact that the Turing machine of is not a really operating device, it is still used as a basic model for explanation such issues as "an algorithm" and "a computational process". It is also used for clarification of the connection between algorithms and computers. The Turing machines are often used for analysis of the computational complexity of algorithms in different control systems. Let us discuss the basic concepts connected with the Turing machines.

The Turing machine (Fig. 10.15) has the finite set of letters s_i, forming the external alphabet. This alphabet is used for encoding information received by the TM and formed by it. The empty sign s_1 is included into the external alphabet. It is used for deleting signs from any cell of the tape.

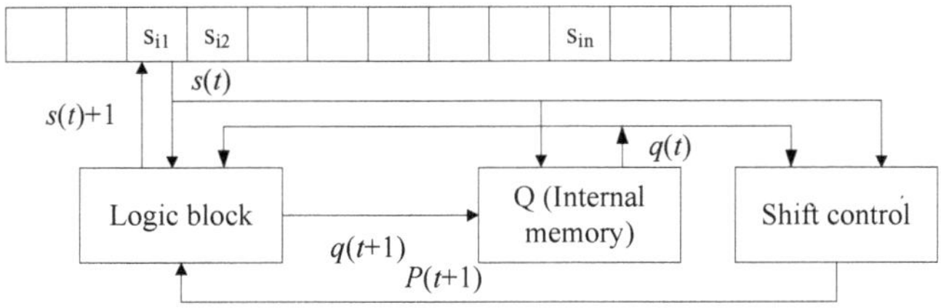

Fig. 10.15 Block diagram of Turing machine

The initial information is determined by the signs which are placed on the tape. Depending on the initial information, two cases are possible.

The first of them is the following. The machine stops after the final number of steps having some information β. The TM forms the signal of stoppage. In this case, the TM is used for transformation of its input alphabet into its output alphabet. In the second case there is no stoppage. In this case the TM cannot be applied for the input alphabet

In any instant of time, only one tape cell is observed. The transition can be made only into some adjacent cell. Three transition operations are possible: shift to the right (*R*), shift to the left (*L*), and no transition (*N*). The transition into some cell needed an arbitrary amount of shift operations is made as a sequence of one-positional transitions. Each step of operation is accompanied with replacement of the observed cell's content s_j by some other sign s_j.

The logic block of TM has the finite amount of states: $\{q_i\}\, i = 1 \ldots m$. The signs $R, L, N,\ q_q, \ldots, q_m$ form the internal alphabet of the TM.

The processed sign $s_j$written into the observed cell is some function of the sign s_j analysed in the current cycle t and the current state $q(t)$of the TM. The same is true for both the next state $q(t+1)$ and the next shift operation $P(t+1)$executed to make a transition. These functions can be represented as the following:$s_j(t+1) = f_1(s_j, q(t))$; $q(t+1) = f_2(s_j, q(t))$; $P(t+1) = f_3(s_j, q(t))$. The program for TM is determined by the triplet $\{s_j, P, q\}_t$.

The example is shown bellow for the program of the inequivalence function calculated by the TM (Table 10.1).

Table 10.1 Program of calculation of inequivalence function

Sign(s_i)	State of TM			
	q_1	q_2	q_3	q_4
0	0, R, q_2	0, N, q_4	1, N, q_4	0, N, q_4
1	1, R, q_3	1, N, q_4	0, N, q_4	1, N, q_4

In the beginning, the TM is in the initial state q_1 connected with the reading the first operand. The state q_4 is the final state; the sign s_j is not changed for the final operation. Let us point out that in this case the TM is applied to the initial information.

Let the program be determined by the Table 10.2. In this case, the TM cannot be applied for the initial information. It is explained by the fact that the sign s_j is always changed by opposite in the state q_4.

Table 10.2 The second variant of the program

Sign(s_i)	State of TM			
	q_1	q_2	q_3	q_4
0	0, R, q_2	0, N, q_4	1, N, q_4	1, N, q_4
1	1, R, q_3	1, N, q_4	0, N, q_4	0, N, q_4

The operating program of a TM is an object with the precise structure. The operating TM is reduced to step-by-step transformation of the current input information into some output data. This transformation is executed according to the program. The input information is determined by the content of the particular cell of the tape; the output data depend on the input sign and the internal state of TM. The operation is terminated when the final configuration is achieved. Obviously, only some determined input information allows obtaining the final configuration. Because of it, the TM can be used for refining knowledge about an algorithm in the given input alphabet. The TM can be used for determining such important properties as function ability to be calculated or to be enumerated recursively and so on. These characteristics give possibility to estimate the chances of planned and developed structures to be implemented. It concerns data flows in the communication channels. For example, the multi-taped TM can be viewed as an adequate model for simulating the multistation access to the some network device. But the numerous researches should be conducted to develop some adequate models using the Turing machine for simulation and analysis of different processes taking place into telecommunication systems.

Recommended Literature

1. Baranov, S.: Logic Synthesis for Control Automata. Cambridge University Press, Cambridge (1994)
2. Barkalov, A., Titarenko, L.: Logic Synthesis for FSM-Based Cotnrol Units. Springer, Heidelberg (2009)
3. Caillaud, B.: Synthesis and control of discrete event systems. Springer, Heidelberg (2002)
4. Cassandras, C., Lafortune, S.: Introduction to Discrete Event Systems. Springer, Heidelberg (2010)

5. David, R., Alla, H.: Discrete, Continuous, and Hybrid Petri Nets. Springer, Heidelberg (2010)
6. Hopcroft, J., Motwani, R., Ullman, J.: Introduction to Automata Theory, Languages, and Computation, 3rd edn. Pearson/Addison Wesley, Reading (2007)
7. Iordache, M., Antsaklis, P.: Supervisory control of concurrent systems: a Petri net structural approach. Springer, Heidelberg (2006)
8. Jensen, K.: Coloured Petri Nets: basic concepts, analysis methods and practical use. Springer, Heidelberg (1996)
9. Linz, P., Kristensen, L.: An introduction to formal languages and automata, 4th edn. Jones and Bartlett (2006)
10. Wang, J.: Timed Petri nets: theory and application. Springer, Heidelberg (1998)

Chapter 11
Business Process Management and Telecommunications

Abstract. The chapter is devoted to management of business processes (BPM). Its role for telecommunications is analysed. In the same time, the role of infocommunications for organization and management of business processes of the general nature is discussed. Some other problems are discussed here, such as the processes of management of activity, peculiarities and main directions of development of infrastructure of information systems, the role of new system technologies of virtualization, service-oriented architecture (SOA), grid-mechanisms, and so on. The restrictions are analysed springing up under the management of business-processes.

11.1 Telecommunications and Business Processes: Short Introduction

The efficient business process management (BPM) is one of the main factors determining the efficiency of any telecommunication company. The same is true for any telecommunication operator. In accordance with the TMN conception, discussed in the section 2.8, the business management layer (BML) is the upper control level. Four other control levels (SML, NHL, EML, and NEL) obey to BML. A business process (or business method) is a collection of related, structured activities or tasks that produce a specific service or product (serve a particular goal) for a particular customer or customers. Obviously, business processes in telecommunications are closely approximated to the business processes from other industrial domains. There are two issues which should be determined in the case of telecommunication business processes. Firstly, it is important to determine both the internal and external structure of the business processes. Secondly, it is necessary to estimate the influence of telecommunication and informational systems on the efficiency of business processes. Moreover, our analysis shows that both the management of infocommunications targeting some business processes and the management of the business processes by themselves are based on the same principles.

V. Popovskij et al.: Control and Adaptation in Telecommunication Systems, LNEE 94, pp. 153–169, 2011.
springerlink.com

Last decades are characterized by radical changing for the place of information technologies (IT) and their influence on business, economics, and life of human society. The information technologies turn into very powerful and sophisticated tools using to support the business activity and the decision making. It can be stated that the normal activity of the human society is impossible without information technologies.

Up-to-day organization of business requires permanent updating information needed for running the business activity. Moreover, this updating should be done in the real time mode. It is achieved due to entry data flows from different sources, as well as queries to different databases, data warehouses, files and so on. These problems concern the problem of receiving data by demand or receiving information-as-a-service (IaaS). The volume of required information is increasing with each year. Now a lot of data are distributed among different corporative systems. The competitive activity requires a really fast data searching and processing. It means that companies need new methods and tools of control (management). The existing tools of business analysis do not offer the required instrument of control. To be successful, a company needs highly productive and reliable technological solutions, as well as it requires some integrated model of activity. Such a model should include the complete presentation of strategic goals, complex of plans, current situation, perspectives, and measures taken for optimizing the company activity. Of course, it should be developed the corporate platform for management of the company efficiency; it include some methodologies, applications, processes and corresponding services.

The statistics shows the number of world servers is doubled each five years, sizes of networks are doubled each two years, and the number of data storage systems is doubled each year. It means that some very efficient methods (and tools) are necessary to control the business activity.

11.2 Content of Control Processes for Activity Efficiency and Peculiarities of Infrastructure

The generalized model of business activity's evaluation includes the discipline of general-corporate management by profits, expenses, assets and finances. Besides, it includes the following eight areas for measuring, result estimation and decision making: finances, marketing, sales, client servicing, product development, manufacturing processes, human resources, telecommunication systems, and information technologies. The components of the generalized model of business process are shown in Fig. 11.1. These components are described by some system of balanced characteristics and metrics. This system allows getting the answer for the following three basic questions: How the current situation is? What are the reasons for this current situation? What should be done? Each activity area have its own methods used for measuring of real characteristics and estimation of their income into the financial and performance targets of efficiency, allocation of responsibilities for the people making decisions.

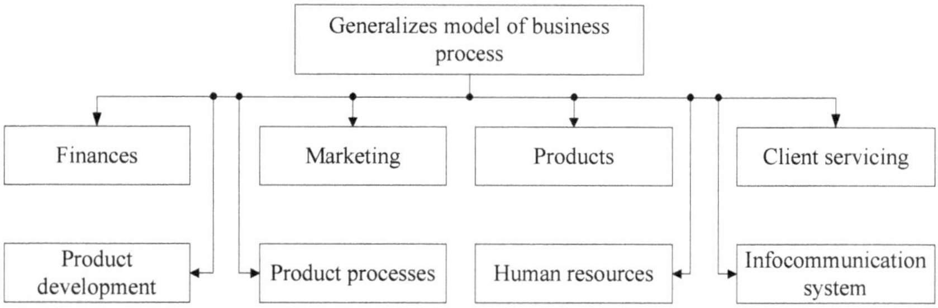

Fig. 11.1 Components of the generalized model of business process

Therefore, the standard closed control cycle with the feedback is formed based on the Watt's principle. This cycle includes the following issues: the planning (formulation of criteria and strategies, coordination of goals, targets and metrics for all layers of management); monitoring of activity (that is monitoring of values for characteristics interested for the management); analysis of results and forming accountability (reports) needed for decision making. The goal of decision making is some correction of goals and forming some control actions. The block diagram of management for a business system is shown in Fig. 11.2.

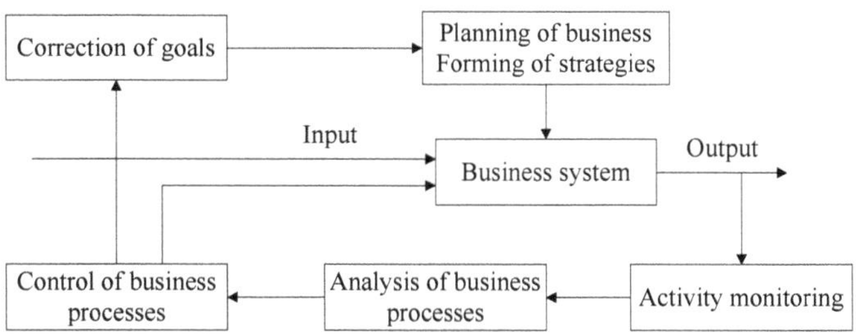

Fig. 11.2 Components of the generalized model of business process

The activity model and strategy specify the principal direction of development and goals (for example, the increase of operating income), key indicators, metrics (for example, the fixed charges or gross receipt), and measurements (for example, the region or finance quarter). The general corporate plans spread on all activity directions and all divisions (departments) of a company. It allows clarification for local criteria and key points. It results in forming the operating plans and distributing resources.

The scenario simulation allows rapid analysis of development variants, follow-up actions for fulfilment the plans; it provides monitoring of actual state of finances, control of resources, relations with clients, and so on. If there are some deviations, the warnings are issued and the causes of deviations are determined using some tools of business analysis. Next, the corrective actions are made and the control cycle makes its new round. To provide such an interrelation among the management processes, the integrated model and control mechanism for business data should exist. Next, the control tools for management of corporate efficiency should be well integrated.

Analysis of business processes used by service-providers (telecoms operators) shows that these processes can be divided by two main groups. The first of them includes the processes determining the development strategy of a given operator, its infrastructure and used products. In other words, these processes determine the life cycle of a process. The second group includes the networking operations executed by a service provider or a telecoms operator. Let us discuss the structure of business processes of telecoms operators.

There are seven groups of processes integrated along the vertical. They are end-to-end processes involving more than one function. These processes are required for supporting the users of communication services, as well as the business management of the operator. The central place here is occupied by the processes of in-service support of networking operations of users. These processes are known as the customer operations processes; they are determined by the abbreviation FAB meaning fulfilment, assurance and billing. Let us point out that the processes of operation support and readiness are separated functionally from the FAB processes. It is specified by the fact that the FAB processes are executed in the real-time mode. Because of it, the FAB processes should be automatized to provide the constant and well-timed supporting customers. The FAB processes have the direct interfaces with customers of communication services; they occupy the central place in the production activity of a telecoms operator.

There are some processes which do not directly connected with immediate support of users. These processes include the strategy of development, the infrastructure lifecycle management, and the product lifecycle management. Their time scale is quite different from the time scale of FAB processes. The years are required for making the infrastructure of telecommunications (for example, building and project construction). But the examination of user accounts before activation of communication session requires only seconds. The life cycle of a telecommunication network is the time from the beginning of infrastructure construction (or introducing some new service) till the demounting of infrastructure and termination of the service providing by the particular telecoms operator.

The base of the strategic evolution of any telecoms operator is represented by the management of infrastructure, as well as management of products and services. Obviously, it is impossible to implement some business plans without plans of the capital construction and plans of development for communication networks. Such issues should be included into the plans as Internet, IP-telephony, videoconferences and so on. All these issues are combined into an integrated strategy of a particular telecoms operator.

11.3 Interaction Model of Business Data Processing Center

The conception of service is a base for the model of interrelation between business and data processing center (DPC) of new generation. This model is shown in Fig. 10.3. All contacts between the business and IT obey to the following pattern: the only thing required from business is the service request to IT, whereas the only thing required from IT is service providing according to SLA. As you remember the acronym SLA stands for service level agreement.

There are three kinds of services: the information service, the infrastructure service, and the application service. The resources from the mutual pool are used for providing each service. This pool is shown as the block "DPC: Standard IT-resources" (Fig. 11.3). There is an automated management system used for control of service configuration, executing reservation of services, providing reservation of necessary resources, as well as monitoring and modification of services. The important feature of the model is the ability of transparent replacement for such components as used IT-resources, service algorithms and even service technologies. The term "transparent" means that these replacements are invisible for the end users of these services.

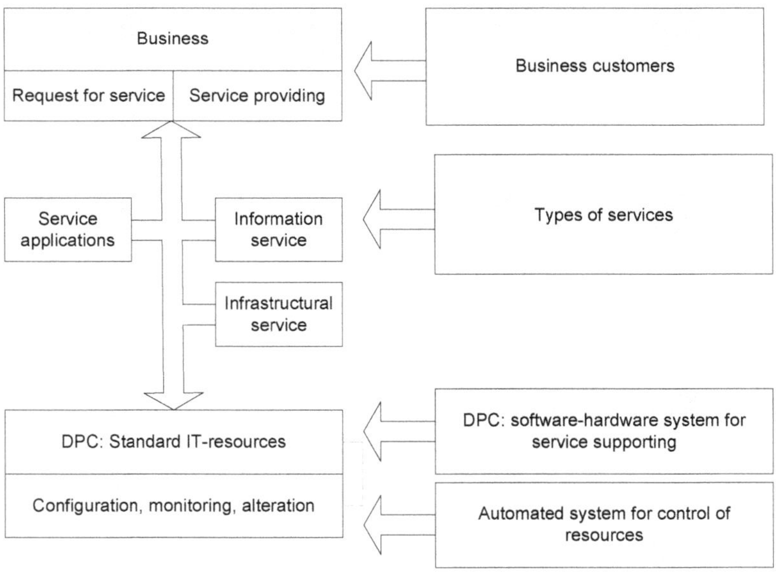

Fig. 11.3 Model of interrelation between business and DPC

Operating experience of computer centres used for supporting such or such business process shows that the variety of organizational, technical and other causes required the creation of centralized DPC. As a rule, these centres are placed apart from the headquarters of a company. They should occupy such places where there is the guaranteed electricity and connections with telecommunication channels having the high throughput.

The data processing centres are viewed as core sets of information infrastructure of up-to-day business. They provide distributing all data flows of a particular company together with their processing, storing and archiving. Each year, the demands to DPC becomes higher and higher. It is connected with technical reasons and with hardening of legislative requirements. It results in complication and higher criticality of the problems connected with management of DPC. Obviously, this problem becomes more and more complex each year.

The special logistical requirements should be observed into modern DPC to provide the highest possible level of accessibility. According to these requirements, such procedures as replacement of equipment, installation of new software versions, changing configuration of cable system should be executed in such a way that they do not affect the current company activity. It means that the paramount attention should be paid to the issues of providing their flexibility, readiness, and simplicity of maintaining. It is necessary to defend investments during the complete time interval; this interval should be preliminary defined.

Initially, the DPCs were created as computing centres having the following goals: providing operating frame computers and servers; storing data using different information tanks (magnetic tapes, disks and so on). To provide it, such issues were the most important as the maintainability and small down-time for application. It was important because just these factors took the critically important meaning for a given enterprise. The following applications can be pointed here: the systems of enterprise resource planning (ERP), the production control systems used by industrial organizations, data bases, some office applications and their operating systems, the systems providing access to telecommunication networks of common use and access to Internet.

The modern infocommunication systems are implemented using seven-layer model of open systems. The highest step of this model is occupied by the application layer; the last step belongs to the physical layer. The physical layer combines all infrastructure necessary for the data transmission, the radio wires or cables, data transmission devices such as hubs, routers, network gates, technologies controlling network elements, networks, and service layers.

The Microsoft Operations Framework (MOF) library includes the methodology for constructing operational processes targeting IT. Also it includes some guidelines for controlling IT-infrastructure. These guidelines give very important, tested instructions for execution of projects, for planning and forming the policy in the IT-area. They also include instructions for application, for management of service and monitoring of IT-services, for controlling configurations and changing. The central chain of these guidelines belongs to the management targeting business problems and constructing efficient IT-activity on behalf of business, as well as the permanent strategic partnership with business. There are a lot of different specific guidelines and manuals. But we think that three their basic principles should be pointed in our book. These principles concern the problem restructuring the IT service management (ITSM). The first principle states that it is necessary to act from the positions of policy system rules (PSR) in restricting process. It means that development and replacement of some part of the system should be coordinated with the tasks and functions of the whole system. The second principle concerns changing into management policy which is built traditionally as some hierarchical structure. This principle advises the rejection from the traditional structure. It is recommended to introduce some new mechanisms supporting monitoring the inter-department ITSM processes. To do it, the process managers are necessary having rights for control and correction of these processes in accordance with existed goals and tasks. In other words, their main task is to improve the operation of IT-service. The third principle is connected with necessity of reorganization inside a company. It is connected with forming a new culture of an organization and style of operation, with stimulating staff to work more efficiently, with taking into account personal and group interests of the IT service staff.

11.4 Basic Lines of Development of Business-Oriented Infocommunication Systems

The basic lines of development of infocommunication systems can be viewed as convergence of the following three new technologies: the virtualization of the network infrastructure, servers of standard architecture, and data processing centres; using the service-oriented architecture (SOA); and using of grid mechanisms.

The convergence of facilities of virtualization, technologies of distributed computing, and service-oriented hardware allows optimizing interactions between business and IT processes. It is connected with the fact that the new technologies allow automatize many processes and modernize the fundamental principles of business functioning. Because of it, many operators take care about integration of the most progressive technologies, such as SOA, tools of virtualization, grid mechanisms, VoIP, multi-core processors, RFID, Wi-MAX, Wi-Fi and so on. The interaction of different initiatives of IT and results of business based on these initiatives is shown in Fig. 11.4.

The virtualization becomes the main tool of efficiency increase for the computing infrastructure of DPC. It allows using fewer resources to solve more problems. It envelopes practically all infrastructure components: from the network and its clients to the storage systems and servers. Further development of virtualization opens perspectives for implementing dynamic infrastructures and cloud computing (CC).

More and more often, the virtualization is considered in the frames of the conception of external and internal clouds. The idea of cloud computing presumes creation, development and usage of computer technologies straight in the Net. It is an organization style of computer systems when the data processing in IT-systems is represented as some services. These services allow data processing without merging into problems of an infrastructure supporting these services. The "Cloud Computing" is a paradigm when information is processed using network computers, whereas the calling processing programs and observing results is provided by a browser.

The CC-infrastructure is provided by the formation of powerful computing centres targeting sharing (collective usage) of hardware with wide use of clusterization

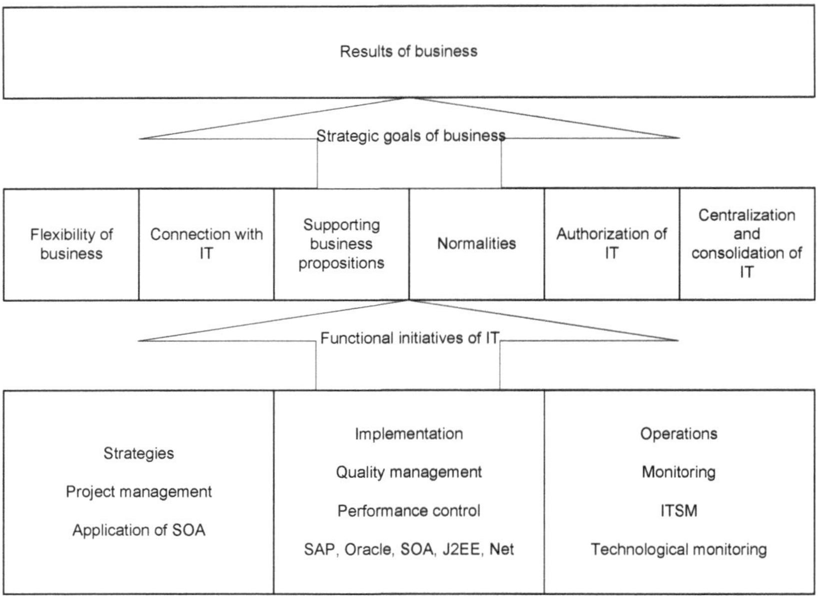

Fig. 11.4 Interaction of IT initiatives and results of business

and process virtualization. The CC-technology presumes the immediate allocation of any resources by a provider according to the request of a user. At the same time, the agreed service level should be provided where the user money should be paid only for actually used resources. The thesis about the repetitive multiple usages of servers is used as an economical explanation of the efficiency of CC-approach. These servers should be developed taking into account the requirements of service architecture or be accessed using the software-as-a-service (SaaS) model.

The on-line postal facilities offered by MSN Hotmail and Google Mail can be viewed as one of the forms of CC. This form is reduced to organization of storing for user data (theoretically unlimited size) and access to them through a browser. The CC-technology allows business to move its data and applications into the Net and to use them in the outsourcing mode. It allows sufficient economy because there is no necessity in buying expensive hardware and software, as well as supporting the permanent availability of applications.

The companies HP, Intel and Yahoo announced creation of the open computing laboratory named the Cloud Computing Test Bed. It targets running researches into the area of Cloud Computing. This laboratory can be viewed as some globally distributed research environment supporting researches targeting software development; its main goal is the procurement of new Web-applications and services. It is planned to produce integrated infrastructural solutions including computing facilities and network products, software for virtualization and data storage systems, safety and management systems. The created solutions allow centralized control of the virtual infrastructure, automated restart of virtual computers, dynamic providing the system resources, and running the complex monitoring of operability of virtual machines.

The examples of projects implemented by Complete company show that virtualization allows significant increase for nonfailure operating time and reliability of servers' providing. The customers get the considerable economy due to more efficient usage of hardware; besides, they do not need additional expenditures of electricity for DPC. At the same time, the process of control is simpler for the service infrastructure.

The virtualization is used for creation so called thin clients (sometimes called lean or slim clients). In this case the computing operations are executed by the server, rather than by a client's personal computer. The thin clients even can have no central processor, operating system, local applications, and data. All functions connected with data processing are displaced into the server. The following tendencies can be predicted for the nearest future: the bench-top systems will turn from the devices into services; the corporative personal computers will be replaced by economical devices of different classes; the shift into corporative environment from fixed access to dynamic access (without lock-on to the fixed workplace); the introduction of the model "service payment for real usage". One of the possible variants of implementation of such clients is shown in Fig.10.5. In this variant, the transition is done from the fixed access to the corporative environment to dynamic access without lock-on to the workplaces.

The following components are shown in Fig. 11.5: the virtualization software (V-Software), the manager of communications (C-manager), the mobile thin client (MTC), and the thin client (TC). The model is transparent enough. The thin clients are connected with data processing resources through some net (Internet, LAN, or WAN) and manager of communications. Next, a client can use practically unlimited powers of virtual machines (VM).

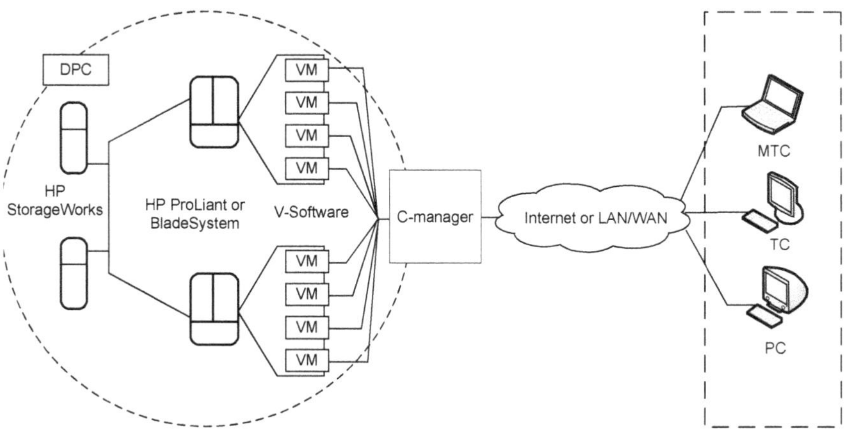

Fig. 11.5 Application of technology of thin clients in dynamic usage of virtual machines

The following objects can be virtualized: the communication channels, the operating systems, the applications, the workplaces of the data storage systems, the data, the servers, the centres of data processing, and the enterprises. The virtualization allows optimizing usage of computing, channel and other resources. For example, the virtualization of the servers with a standard architecture can be made on the base of the complete virtualization with binary translation on the base of paravirtualization and virtualization of hardware resources. In computing, the paravirtualization is a virtualization technique that present into a software interface to virtual machines that is similar but not identical to that of the underlying hardware. It assumes existence of the hardware with very high performance which can be used to solve so called guest's tasks. Let us point out that solution of these tasks into virtual environment is more time-consuming than in the case of paravirtualization.

There are two groups of methods called respectively micro-virtualization and macro-virtualization. In the first case, the shared hardware (a processor or service equipment) is divided by the variety of environments (virtual machines). Each virtual machine has its own operating system (for example, Linux) and applications. The macro-virtualization is provided due usage of grid-mechanisms. Let us discuss the grid-technologies more thoroughly.

The grid computing means combining different computing resources to achieve some common goal. Thanks to grid-approach, the available resources (computing powers, storage resources, network equipment) are not fixed for some specific device or node. They can be used inside the distributed and dynamic information infrastructure and they are combined into a mutual pool using the corporative grid-technologies. The combining intermediate software of grid-networks with virtual machines results into a grid-network of virtual machines, where each of them is connected with different type of resources.

There are significant differences in the grid- and cloud-architectures. It is connected with the fact that these architectures were developed on the base of different preconditions. The grid-architectures were created to use efficiently some expensive distributed computing resources and to make these resources dynamic and homogenous. Because of it, the grid-architectures focus on the integration of already existed resources, including hardware and software, operating systems and local tools used for providing safety and control. As a result, a virtual enterprise is appeared and only participants of this enterprise can consume its resources. This enterprise's existence is supported by five layers of protocols, tools and servers. This organization is shown in Fig. 11.6a.

The infrastructural or fabric layer is the lower one. It combines computers, storage systems, networks, repositories of codes. Next is the connectivity layer, where the specific protocols of communications are determined. The resource layer provides assignment of resources, tools for control of resources, distribution of resources among the users, and billing. The collective layer supplements the resource layer, giving opportunity of using some complexes of resources. At last, the application layer is used for supporting applications.

In contrast, the cloud-architecture is open for access through the Net, which is, of course, much wider than any grid-network. There are standard protocols for access to pools of computing resources and data storage pools. There are such protocols as WSDL and SOAP; at the same time such advanced technologies as, for example, Web 2.0 use protocols such as REST, RSS, AJAX. The existed grid-architectures also provide access to cloud-resources. The cloud-protocols can be divided by four layers (Fig. 11.6b). The fabric layer includes some "raw" computing resources, such as servers, storage systems, and networks. The unified resource layer includes the same resources, but they are represented in some abstract form. These abstract resources can be assigned to users or to upper level as virtualized servers, clusters, file systems, and database management systems (DBMS). The platform layer adds some specialized instruments for connection of software and servers above the universal resources. It forms the environment for development and implementation of application. The application layer contains applications executed into the cloud environment.

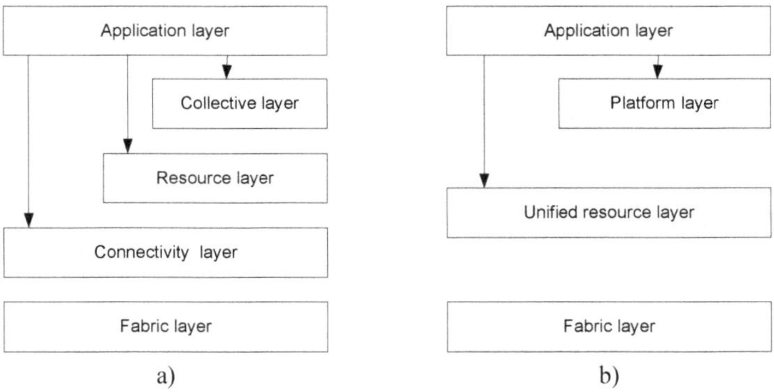

Fig. 11.6 Comparison architectures grid (a) and cloud computing (b)

From the user's point of view the servers of clouds can be divided by three main levels: IaaS (Infrastructure-as-a-Service), PaaS (Platform-as-a-Service), and SaaS (Software-as-a-Service). Besides, such services can be accessible and required as: IdaaS (Identification-as-Service), HPCaaS (High Performance-as-a-Service), and other high-performance servers. Extrapolating evolution of IT, some experts predict the era of EaaS meaning "Everything-as-a-Service".

The grid-technologies can be used in a variety of application domains. But all of them have one mutual property: the packet approach. It means that a complex of software tools is selected, compiled and placed taking into account the specific platform and user's payment abilities. A packet can include both universal and specific tools (determined by the needs of a specific project. But the Achilles hill of the packet approach is lack of some universal architecture of products developed by different users. It can become apparent during the compilation of software platform with tools of different producers.

Let us discuss the grid-architecture represented by the cluster file system Lustre of Sun Microsystems. The Lustre allows organizing the computing configuration with required power. The main intention of creation the Lustre was in creation of a file system having tens of thousands nodes having together around a few petabytes (PB). It is worth mentioning that 1PB is equal to 1024 terabytes or (2 in power 50) bytes. This scaled system includes three basic components:

1. A server of metadata with either active or passive reservation which is responsible for the structures of catalogues, attributes and file names.
2. A storing server which includes a few portions of data ("storage goals"). Each portition can be up to eight terabytes.
3. A client uses the server of metadata to find a required file. Next, the client blockades some displacement diapason inside the file and uses the storing servers to make modification of necessary data.

In each specific case, the construction of a grid-system is a nontrivial task; it requires an analysis of initial conditions, as well as possible variants for applying software-hardware tools from different manufacturers. Now, let us discuss the methods used into service-oriented architectures.

A service-oriented architecture (SOA) is a complex of products providing the business process management. The following products can be included into SOA:

1. The products of enterprise service bus (ESB) type providing data transfer among different servers.
2. The systems of administration of application development called the systems of design – time governance.
3. The systems for control of the operating environment of servers called the runtime management.
4. The security gateways providing safety of data transferring.

The SOA envelopes the complete totality of services provided by a network. It includes the servers used by applications, the services used for communications and interrelations of human beings and information systems, and the servers supporting the infrastructure of a particular network. For each layer, the components of SOA execute some definite logical functions; the architecture is modular and open. Companies-users can define these functions taking into account their own specifics. Due to compacted integration of layers, it is simpler to organize intercommunications among applications and network services. At the same time, the pre-conditions are created for increasing performance and expanding the functional facilities of the total network.

The SOA allows operative execution of business intelligence (analytics), as well as it can separate individual business functions and provide them as services which can be multiple used into end-to-end business processes. It is known that the business intelligence (BI) is one of the most difficult tasks belonging to the category of "complex events' processing". It is provided by such thriving technologies as the following:

1. The interactive virtualization or virtual data representation. The tools of virtualization should be transformed and adapted. It means that a user gets tools for input-output control.

2. Analytical tools. These tools are capable to operate with data located into random access memory.
3. Integration of BI with technologies of corporative search.
4. The software as a service.
5. The service oriented architecture.

The results of research prove the advantages of systems assembled from some loosely coupled modules using the SOA. But let us point out that this approach is a compromise (as any engineering decision). The negative feature of SOA is the reverse of decentralization. It means that application of SOA leads to the loss of the unified system of data storage; the data are now distributed among the different services. This fact has no essential value for the majority of transactional systems. But the decentralization in the data storing is a critical issue for BI-systems where it is necessary to analyse the data (including historical ones). It results in the existence of the gap between SOA and BI. So, now there is an actual task of efficient combining tools of SOA and BI.

One of the perspective solutions for this problem is the following approach. It is necessary to track the states of servers (in some specific way) and take out the changing in data. But the volume of these data can be huge and it can result into congestion of data transferring channels and storage systems. Of course, it is possible to introduce some limited filtration. For example, the data can be scanned with some determined on-off time duty ratio. But it can lead to loss of some important events.

The constructive approach from this situation can be the solution based on the model of compulsory (push) delivery. This method becomes more and more popular in the pair with the architectural approach called event-driven architecture (EDA). The essence of this method consists into usage of a server to manifest events taking places in this server. It could be done either periodically or when some event occurs. This approach provides possibilities for implementing such technologies as ESP (Event Stream Processing) and CEP (Complex Event Processing). Of course, the EDA-approach can be implemented using some tools different from SOA, but the paradigm "EDA built on SOA" solves all problems of BI. If the event stream processing is executed, then the BI components gather required data, execute their filtration, and store them into data storing systems or send them to data marts.

As a result of this convergence, the integrative processes of enterprises front the new level. The business processes connect with IT-processes; it provides the high level of dynamic for these processes. The further integrative steps lead to possibility of optimization for all stages of operating process where such chains participate as partners, providers, and clients. As a result, the boundaries between separate enterprises are washed away and the global virtual ecosystem is created. The tremendous perspectives are opened for this ecosystem; it provides active implementation for new solutions and, as a result, to growth of business efficiency.

11.5 Directions for Implementation of Business Process Management

The problems of BPM are in the focus of scientific and practical attention in IT-analysis. The main goal of BPM is the providing strategic preferences to the main business. The rational use of IT allows increasing efficiency of business processes due to decrease of their resource-intensiveness, execution time, growth of their staff's working efficiency, cost cutting, and increasing the competitiveness of business integrally.

The processes taking places into telecommunication systems are the important component of the integrated complex created by business and IT-processes. The TCS can be viewed as the stock base of IT-systems.

It is well known that both business and IT systems represent very complex logistical formations. But a lot of processes are automated for these systems. For example, it could be atomized such procedures as monitoring of service management applications in SOA, service-oriented management, control of interfaces and transactions, and so on. The standards are implemented using for management of distributed IT-environment. For example, three companies (CA, IBM and Talking Blocks) developed the standard WSDW (Web Services Distributed Management), as well as two additional specifications: MUWS (Management Using Web Services) and MOWS (Management of Web Services). These tools determine the presentation of control interfaces for arbitrary IT-resources as Web-services, as well as the control by these interfaces. The IBM promotes the ideas of computing systems with automated self-control; it is so called autonomic computing. At the same time, it is impossible to make the really important business decisions without participation of a human being. So, PMD plays important role into control of the complex combining business processes and IT-systems.

The situation inside TCS can be viewed as a quite different. As a rule, human beings make decisions here; there are some on-duty shifts of operators and so on. But a lot of TCS-processes are routine and multiple repetitive. Because of it, such devices as automated commutators are wide used, as well as some other automated solutions. A lot of software is developed for providing the automated access control, communications, service providing, restructuring of networks. But new and new tolls are required to increase the efficiency of TCS. It means that it is needed modernization and improving of existed algorithms and development of new ones.

The solutions are more efficient if they target some wide class of situations, rather than some specific situation. It means that the control procedures should be optimized taking into account their stochastic nature, as well as some specific operating conditions. These conditions include operation with delays into control loop, existence of statistic and functional dependences between different values, processes, and fields, as well as influence of different restrictions (constraints) into channels and network structures.

But there is one fact more important that all these restrictions. It is the requirement of a system approach. It means that the fact should be taken into account that

TCS is an embedded part of general-system complex including business and IT-processes. Therefore the processes taking places into TCS should be subordinated to the general ideology of this complex's operating. The general rules of system policy should be used to solve this problem.

It means that there is a necessity in formal adaptation of TCS to requirements of business processes, such as simplicity of access, confidentiality, implementation of FMS conception (fixed mobile communication service) and so on. But maybe more important is account of underlying requirements leading to correction or transformation of telecommunication technologies for all seven layers of an open system. The experts think that convergence of business processes, IT-processes and telecommunication systems will result into significant changing of traffic dynamic, growth of the size of ordering information, as well as its value.

11.6 Constraints into Business Process Management

The idea of creation of distributed information-computing environment has appeared due to existence of some specific constraints in the computing medium. But there are also some constraints specific for telecommunications. For example, the throughput of communication lines is limited. It can be expressed by the following equation

$$C = \Delta F \log_2(1 + P_s/P_n). \tag{11.1}$$

The actual limit is restricted not only by the bandwidth ΔF and signal/noise ratio. Different network elements (access points, commutators, routers, buffers, and so on) constrain the throughput even more, than the factors from equation (11.1). It means, there is a conflict between needs of users and facilities of a network. Obviously, there are conditions when combination of restrictions and requirements give the maximum effect for both telecommunication and IT-systems.

The next constraint for organizing distributed computing is the data delays. Let some packet appear on the input of a network in the instant of time T_{in} and let it appear on the network output in the instant of time T_{out}. The general network delay (latency time) T_{Σ} for an information packet is determined as the difference between T_{out} and T_{in}. It can be described by the following equation:

$$T_{\Sigma} = T_{out} - T_{in}. \tag{11.2}$$

This delay includes the following components: a delay in the transmission along the communication channel; a delay connected with processing of the packet; a delay connected with waiting in different queues.

The data delays affect the computing processes, but their influence is especially substantial for implementation of control algorithms. In the tasks of situational control these delays are not so critical because of other delays connected with nonoptimal nature of situational control (influence of a human factor). But in the case of optimal control, the influence of delays is significant and it mostly determines the general efficiency of the task to be solved. In common case, the states of a dynamic system with the delay $\Delta t = \tau$ and control action $u(t)$ is determined by the following equation:

$$dx(t)/dt = F\{t, x(t), x(t-\tau), u(t)\}. \quad (11.3)$$

A lot of researches are devoted to the problem of finding control actions $u(t)$ in the systems with delays. There are two kinds of control algorithms: with delay by state and with delay in the observation channel. The control algorithm with delay by state is determined by two equations. The first of them is the following functional-differential equation:

$$dx(t)/dt = A_1(t)x(t) + A_2(t)x(t-\tau) + B(t)u(t). \quad (11.4)$$

The second equation is a watch equation:

$$y(t) = H(t)x(t) + \xi(t). \quad (11.5)$$

In 11.5 the function $\xi(t)$ determines the noise in the observation channel.

The control algorithm with delay in the observation channel includes the classical state equation:

$$dx(t)/dt = A(t)x(t) + B(t)u(t). \quad (11.6)$$

In the case the watch equation is the following one:

$$y(t) = H(t)x(t-\tau) + \xi(t). \quad (11.7)$$

The observations play the determinative role into the control tasks. Obviously, a system is not controlled, if it is not observable. There are different mechanisms providing observing and monitoring in TCS. For example, the SNMP protocol is used the TCP/IP networks. This protocol provides control and monitoring of network elements, configurations, performance, safety, and gathering the statistics. The RMON protocol is an extension of SNMP. The RMON protocol provides gathering information about events and devices where the hardware-software agent is installed; it also delivers data about the traffic characteristics among the network devices. These data are used for making decisions about the changing in the structure or operating modes of some elements, or network fragments, or of the complete system. In short, these data are used for management of TCS.

Let us point out that it is very important to classify correctly the situation with delays. If there is the delay by state, then we deal with the task with afteraction. Its solution is rather cumbersome. If there is the delay in the observation channel, then we should solve the classical task using both the decomposition theorem and state predictions for the time τ instead of the estimates of states.

Therefore, the telecommunication systems are on the threshold of significant technological changing connected with perspectives of development of business processes and IT-technologies. The today's methods and algorithms of control based on situational approach do not satisfy the requirements of tomorrow's telecommunication systems. These methods and algorithms can and have to be significantly modernized for their optimization and correspondence to the processes to be serviced.

Recommended Literature

1. Bell, M.: SOA Modeling Patterns for Service-Oriented Discovery and Analysis. Wiley, Chichester (2010)
2. Erl, T.: Service-Oriented Architecture (SOA): Concepts, Technology, and Design. Prentice-Hall, Englewood Cliffs (2005)
3. Erl, T.: SOA Principles of Service Design. Prentice-Hall, Englewood Cliffs (2007)
4. Ferguson, D., Stockton, M.: Service-oriented architecture: Programming model and product architecture. IBM Systems Journal 44 (2005)
5. Josuttis, N.: SOA in Practice: The Art of Distributed System Design (Theory in Practice). O'Reilly Media, Sebastopol (2007)
6. Strosnider, J.K., Nandi, P., Kumaran, S., Ghosh, S., Arsnajani, A.: Model-driven synthesis of SOA solutions. IBM Systems Journal 47 (2008)
7. Valipour, H., AmirZafari, B., Maleki, N., Daneshpour, K.: A Brief Survey of Software Architecture Concepts and Service Oriented Architecture. In: Proceedings of 2nd IEEE InternationalConference on Computer Science and Information Technology (2010)

Chapter 12
Conclusion

The important goal of our book is an attempt to fill up the gap taking place into existed systematic information about telecommunication systems. Mostly, this information can be found into different scientific journals. We would be glad if our book induces to discussion of new scientific, technical and technological solutions. Obviously, there are very multifarious problems connected with scientific investigations of telecommunication systems. At the same time, we can point out the most important problem in this field. It is a problem of creation of the common theory of telecommunication systems.

We think that the future building of the theory of telecommunication systems will be built on the base of the system approach. In this case the problems of synthesis and analysis in telecommunications will be solved using the apparatus of the system theory. Now, they are solved on the base of technological improving.

Unfortunately, the adopted conception of the system policy rules (SPR) is not always appropriate for solving contemporary problems. There are some causes for this situation. The main reason is that the development of telecommunication systems moves along the road of the search for rational technologies. Obviously, there are practically infinite amount of these technologies; it is very difficult to compare them using some objective metrics. At the same time the existing practice of secrecy for the contents of technologies stems the tide of implementing SPR.

It is very important to use the system approach for analysis of the following pyramid taking place in our society. It is the pyramid including such components as the policy, the business, the information systems, and the telecommunication systems. Obviously, the first component of this pyramid cannot be formalized. But business, and especially business processes, is mostly forecasted. At the same time, all processes taking places into telecommunication systems can be formalized. Therefore, successfulness and efficiency of the business process management depends directly on the level and service quality of required infocommunication services.

It is necessary to draw the readers' attention on some very important problems which are not well presented into available scientific and technical literature. The solutions of these problems will contribute into further progress in telecommunications.

V. Popovskij et al.: Control and Adaptation in Telecommunication Systems, LNEE 94, pp. 171–172, 2011.
springerlink.com

First of all, let us point such problems as development and application of adequate mathematical dynamic models. Such phenomena as randomness of the traffic and other random factors require using stochastic differential models as adequate models of telecommunication systems. Moreover, the solutions obtained for stochastic models describe the classes of situations, whereas a solution of a deterministic model concerns only single specific situation. Huge amount of scientific investigations is devoted to setting and solution of these problems. At the same time, there is a wide class of practical applications where these investigations were not conducted thoroughly (or at all). The multidimensional stochastic state models of random processes can be viewed as one of the most important examples of uninvestigated problems. The transition from the multidimensional space is executed in the trivial way into the mathematics. But it is necessary to execute the nontrivial interpretation of inter-element connections to move from the known one-dimensional cybernetics model to the multidimensional model. As we know a system is a set of interrelated elements and namely these interrelations give some new properties to the final system. Therefore, such properties as emergence and integrity of a system are determined just by levels of interrelations among its components. Because of it some questions arise in the analysis of a particular TCS. For example, there is a question about system properties of such or such TCS. Or it could be asked whether there is some connection between the properties of emergence (and integrity) and qualitative characteristics of TCS. These questions wait to be answered.

There is one more very important problem connected with mathematical simulation. It is known from the system theory that only one property of a system (functional or structural) is, as a rule, represented in its models. It makes narrower the generality of obtained solutions. So, it is necessary to develop models combining both functional and structural properties of systems. It is necessary to find some operator, similar to the Fourier operator in the theory of signals, providing easy transitions between functional and structural system properties. It would be a serious breakthrough in the system theory. If such an operator does not exist, it is very important to find mathematical methods supporting simultaneous execution of analysis, synthesis, optimization, and other important procedures into spaces of functional and structural properties. We think these solutions should be looking for among tensor solutions using deakoptical tensor models.

Modern tendencies into management of business processes implemented on the base of infocommunication services are directed on virtualization of network procedures, development of cloud computing technologies, grid-technologies, service-oriented architectures, and similar things. All this requires corresponding corrections into training and professional development of personnel for telecommunication systems.

The authors will be grateful for any constructive criticism of this book. We are ready for dialog and cooperation.

Index

GPSR Compliance
The European Union's (EU) General Product Safety Regulation (GPSR) is a set of rules that requires consumer products to be safe and our obligations to ensure this.

If you have any concerns about our products, you can contact us on

ProductSafety@springernature.com

In case Publisher is established outside the EU, the EU authorized representative is:

Springer Nature Customer Service Center GmbH
Europaplatz 3
69115 Heidelberg, Germany

www.ingramcontent.com/pod-product-compliance
Ingram Content Group UK Ltd.
Pitfield, Milton Keynes, MK11 3LW, UK
UKHW022342280826
474907UK00003B/36
* 9 7 8 3 6 4 2 2 0 6 1 3 9 *